Reinventing Mechatronics

Xiu-Tian Yan · David Bradley ·
David Russell · Philip Moore
Editors

Reinventing Mechatronics

Developing Future Directions for Mechatronics

Springer

Editors
Xiu-Tian Yan
Design, Manufacture and Engineering Management
University of Strathclyde
Glasgow, UK

David Russell
Malvern, PA, USA

David Bradley
Dundee, UK

Philip Moore
Department of Automation Engineering
School of Engineering Science
University of Skövde
Skövde, Sweden

ISBN 978-3-030-29130-3 ISBN 978-3-030-29131-0 (eBook)
https://doi.org/10.1007/978-3-030-29131-0

This Springer imprint is published by the registered company Springer Nature Switzerland AG
The registered company address is: Gewerbestrasse 11, 6330 Cham, Switzerland

Preface

Mechatronics 2018

The concept of mechatronic systems was first used in Japan in the 1960s by Tetsuro Mori to reflect the emerging role of electronic components in the control and operation of what had previously been inherently mechanical systems. Since then, mechatronics has been at the centre of fascinating technological developments and trends in the modern world and now encompasses and embodies Cloud-based technologies, Internet of Things (IoT) and Cyber-physical systems.

The Mechatronics Forum Conference was the first conference series worldwide specialising in the discipline and has the longest standing programme of international events covering thirty years. *Reinventing Mechatronics* presents selected papers from *Mechatronics 2018*, the 16th Mechatronics Forum International Conference held at the University of Strathclyde in Glasgow from 19 to 21 September 2018, organised by Prof. Xui-Tian Yan and hosted in the Technology and Innovation Centre building.

The conference was held over three days incorporating thirty-five presentations of advanced mechatronics-related research from leading international researchers from the USA, Asia, the Middle East, Europe and the UK. The conference included two invited plenary speakers in Dr. Harald Wild, Professor of Control and Mechanics at the Swiss Universities of Applied Sciences and CEO of Wild Controls and Stephen Sanders, Strategic Director for Remote Operations at Veolia Nuclear Solutions. Furthermore, a special session was held at the conference on '*What is the Future of Mechatronics in Manufacturing?*' where invited guest speakers included Alan Smith, General Manager, Beckhoff and Dr. Erfu Yang, University of Strathclyde.

Conference sessions covered a diverse range of mechatronics-related topics and applications broadly related to the theme of Reinventing Mechatronics including:

- Reinventing Mechatronics or How to Deal with Industry 4.0
- IoT and Artificial Intelligence—A New Challenge?
- What is Mechatronics and Future Mechatronics

- Reinventing Mechatronics for Extreme Environments
- Medical Systems
- Autonomous Vehicles and Drones
- Simulation and Modelling
- Sensors and Actuators
- Manufacturing
- Control.

The Mechatronics Forum

The Mechatronics Forum was formed as an interest group in the UK and is currently sponsored by and part of the Institution of Mechanical Engineers (IMechE), London. Founded in 1990, the Forum has enabled professional engineers from industry and academia from around the world to share innovative ideas in this field and act as body to champion the discipline. The Mechatronics Forum came into existence at a meeting held at IMechE's London headquarters on 30 October 1990.

It was the first organisation in the Western world that recognised the importance of mechatronics and began to promote it. Although the word mechatronics has been around since the late 1960s, it was only in the early 1990s that it was used to any great extent in the UK and Europe. However during the 1990s, with the activities of the Mechatronics Forum, the term mechatronics and the engineering discipline that it encompasses became widely recognised.

Mechatronics today extends beyond the integration of mechanical, electronic and computer engineering. Many engineers now see it as embracing a wider range of engineering activities, from design through manufacture to the market place. Hence, they regard mechatronics as a major influence in pulling together and integrating the many aspects of engineering which with increased specialisation have tended to push apart from each other during past decades. It was in an attempt to solve this increasingly challenging problem that the Mechatronics Forum was conceived as a first step towards the building of bridges between the many technologies, philosophies and disciplines which comprise mechatronics and the professional institutions that are committed to their own particular specialised subjects. In the UK, engineering institutions are important in sharing technical subjects between professionals in industry and academia. They accredit undergraduate and postgraduate courses as suitable for covering the academic components of a Chartered Engineer's development, and they grant Chartered status to those whose careers show sufficient engineering responsibility and understanding to be leaders in their field.

The Mechatronics Forum for its first ten years was supported under inter-institutional arrangements, with secretarial and administrative services provided alternately by the Institution of Mechanical Engineers (IMechE) and the Institution of Engineering and Technology (IET). Following this, the Forum has been supported by the IMechE and linked with the Mechatronics, Informatics and Control Group (MICG).

Mechatronics 2020 will be held in the University of Leuven in Bruges, Belgium.
The following lists Prestige Lecturer to date:

1995 *The Role of Xero-Mechatronics in New Product Development*
Dr. John F. Elter, Xerox Corporation

1996 *Advances in Mechatronics: The Finnish Perspective*
Vesa Salminan, Fimet, Finland

1997 *The Industrial Benefits of Mechatronics: The Dutch Experience*
Professor Job van Amerongen, University of Twente, The Netherlands

1998 *Virtual Worlds—Real Applications: Industrial and Commercial Developments in the UK*
Professor Bob Stone, University of Birmingham

2000 *Mechatronics Solutions for Industry*
Professor Rolf Isserman, University of Darmstadt, Germany

2001 *Intelligent Mechatronics: Where to go*?
Professor Toshio Fukuda, Nanyang University, Japan

2003 *Bionics: New Human Engineered Therapeutic Approaches to Disorders of the Nervous System*
Professor Richard Norman, University of Utah, USA

2004 *GM's Approach to Eliminating Complexity and Making the Business More Successful*
Dr. Jeffrey D. Tew, General Motors R&D Center

2005 *Mechatronic Design Challenges in Space Robotics*
Dr. Cock Heemskerk and Dr. Marcel Ellenbroek, Dutch Space, The Netherlands

2006 *Cyborg Intelligence: Linking Human and Machine Brains*
Professor Kevin Warwick, University of Reading

2007 *Iterative Learning Control—From Hilbert Space to Robotics to Healthcare Engineering*
Professor Eric Rogers, University of Southampton

2008 *World Water Speed Record Challenge—The Quicksilver Project*
Nigel McKnight, Team Leader and Driver, Quicksilver (WSR) Ltd.

2009 *Meeting the Challenges and Opportunities of Sustainability Through Mechatronics Product Development*
Professor Tim McAloone, Technical University of Denmark, Denmark

2010 *I'm a Control Engineer: Ask Me What I Do*?
Professor Ian Postlewaithe, University of Northumbria

2011 *Sports Technology: The Role Of Engineering In Advancing Sport*
Dr. Andy Harland, Loughborough University

2013 *Modelling, Control and Optimisation of Hybrid Vehicles*
Lino Guzzella, ETH Swiss Federal Institute of Technology, Zurich, Switzerland

2015 *The Past, Present and Future of Mechatronics: A Personal Perspective*
Professor David Bradley, Abertay University

2017 *A Review of Current European Space Robotics Research*
Professor Xui T. Yan, University of Strathclyde

In addition to the Biennial International Conference series and the Prestige Lectures, the Forum has supported and facilitated many other activities related to mechatronics and its wider promotion including:

- Technical visits to industrial and academic organisations where these are open to members.
- Technical seminars and workshops focused on particular aspects of the discipline.
- Mechatronics Student of the Year Award, a competition open to final-year degree students and master's level students based on their final project dissertation.

To conclude, the Mechatronics Forum is part of the Mechatronics, Informatics and Control Group of the Institution of Mechanical Engineers. It has had an important role in popularising mechatronics in the UK and beyond, focusing on educational issues and promoting linkages between industry and academia, and seeking ways of bringing together all those interested in mechatronics. Reinventing Mechatronics has taken a selection of the most significant papers and topical areas from the most recent International Conference-MECHATRONICS 2018—*Reinventing Mechatronics* and presents them in extended form as chapters in their own right.

Prof. Philip Moore, Ph.D., FIET., C.Eng.
Department of Automation Engineering
School of Engineering Science
University of Skövde
Skövde, Sweden
e-mail: philip.moore@his.se

Acknowledgements

The editors of the book, as representatives of the Mechatronics Forum, would like to thank the Mechatronics, Informatics and Control Group (MICG) of the Institution of Mechanical Engineers (IMechE) and the Robotics and Mechatronics Technical and Professional Network of the Institution of Engineering and Technology (IET) for their support of *MECHATRONICS 2018—Reinventing Mechatronics*, the 16th Mechatronics Forum Biennial International Conference held at the University of Strathclyde, Glasgow, Scotland, UK, in September 2018.

The editors would also like to thank the following researchers for their invaluable support in organising the conference: Dr. Karen Donaldson, Dr. Silvio Cocuzza, Ms. Youhua Li, Mr. Scott Brady and Mr. Cong Niu.

Contents

Reinventing Mechatronics 1
Xiu-Tian Yan and David Bradley

Reinventing Mechatronics—A Personal Perspective 11
Stephen Sanders

From Mechatronics to the Cloud 17
David Bradley, David Russell, Peter Hehenberger, Jorge Azorin-Lopez, Steve Watt and Christopher Milne

Path Planning for Semi-autonomous Agricultural Vehicles 35
Markus Pichler-Scheder, Reinhard Ritter, Christian Lindinger, Robert Amerstorfer and Roland Edelbauer

The *AgriRover*: A Reinvented Mechatronic Platform from Space Robotics for Precision Farming 55
Xiu-Tian Yan, Alessandro Bianco, Cong Niu, Roberto Palazzetti, Gwenole Henry, Youhua Li, Wayne Tubby, Aron Kisdi, Rain Irshad, Stephen Sanders and Robin Scott

Assistive Gait Wearable Robots—From the Laboratory to the Real Environment 75
Alireza Abouhossein, Uriel Martinez-Hernandez, Mohammed I. Awad, Imran Mahmood, Derya Yilmaz and Abbas A. Dehghani-Sanij

A High Fidelity Driving Simulation Platform for the Development and Validation of Advanced Driver Assistance Systems 93
Marco Grottoli, Anne van der Heide and Yves Lemmens

Advancing Assembly Through Human-Robot Collaboration: Framework and Implementation 111
Abdullah Mohammed and Lihui Wang

Tracking Control for a Fully-Actuated UAV 127
Matthias Konz, David Kastelan and Joachim Rudolph

Simplified Manufacturing of Machine Tools Utilising Mechatronic Solutions on the Example of the Experimental Machine MAX 145
Steffen Ihlenfeldt, Jens Müller, Marcel Merx, Matthias Kraft and Christoph Peukert

Simulation in the Design of Machine Tools 163
Daniel Spescha, Sascha Weikert and Konrad Wegener

Reinventing Mechatronics—Final Thoughts 179
David Russell

About the Editors

Xiu-Tian Yan Professor of Mechatronic Systems Technology, University of Strathclyde, UK.

He received his Ph.D. from Loughborough University of Technology, UK, in 1992 and is Chartered Engineer, Fellow of the Institution of Engineering and Technology (FIET) and Fellow of the Institution of Mechanical Engineers (FIMechE). He is currently Professor in mechatronic systems technology and Director of the Space Mechatronic Systems Technology Laboratory (SMeSTech) in the Department of Design, Manufacture and Engineering Management at the University of Strathclyde. He has chaired other Robotics and Mechatronics Technical Professional Network of the IET and is Technical Editor of IEEE/ASME Transactions on Mechatronics.

His research interests lie within the broad area of mechatronic system design, including robotics, multi-perspective mechatronic system modelling, design and simulation along with the use of AI techniques. These interests have been converted into a wide portfolio of mechatronic research projects including the development and application of mechatronic design process models, the modelling and simulation of novel mechatronic systems and techniques for the prototyping of these systems for purposes of validation. In recent years, his research has focused on space robotics through three projects funded under the Horizon 2020 Strategic Research Cluster on Space Robotics Technologies (H2020-SPACE) and their terrestrial applications in agriculture and advanced manufacturing.

He has published over 240 technical papers in major international journals and at conferences and has authored and edited 7 books in the fields. He has also organised and chaired four international conferences.

In addition, he is Invited Professor or Visiting Professor at four international universities, and in 2017 he delivered the Mechatronics Forum Prestige Lecture *A review of current European Space Robotics Research* and has lectured at universities in France, Japan and China along with various international research institutions.

He is the recipient of Judges Commendation Best KTP Partnership (Scotland) 2010 Award and several best paper awards and has been Key Team Member involved in several industrial awards for mechatronic products or systems.

David Bradley Professor Emeritus, Abertay University, UK.

He was awarded his B.Tech. and Ph.D. degrees at University of Bradford in 1968 and 1972 and, after a period as Teaching Assistant at the University of Toronto, joined Lancaster University Engineering Department in 1972.

While at Lancaster, he was one of the academic team members responsible for introducing in the mid-1980s some of the first master's programmes in mechatronics to be established in the UK and, with Prof. Jim Hewitt and Prof. Jack Dinsdale, was Founder Member of the UK Mechatronic Forum, now the Mechatronics Forum, of which he remains Honorary Life President.

Following periods on sabbatical leave with New Zealand Electricity and the Technical University of Denmark, he moved to take up a chair at Bangor University in 1995 and then in 1998 to Abertay University, where he is now Professor Emeritus.

Chartered Engineer and Fellow of the Institution of Engineering and Technology (FIET), his teaching and research have focused on the design and implementation of complex, and essentially mechatronic, systems including applications such as robotic excavation, the modelling and fuzzy control of hydroelectric plant and e-health technologies. He is currently working with research groups in the UK and Europe on projects such as intelligent prostheses, the design of mechatronic and cyber-physical systems and user privacy. He has been Lecturer or invited speaker at universities and conferences in locations as diverse as Colombia, Malaysia, China and Japan as well as in Europe, Canada and the USA.

He is Author, Co-author or Contributing Editor for 8 books, including 5 on mechatronics, over 50 journal publications and more than 120 conference publications, and in 2015 he delivered the Mechatronics Forum Prestige Lecture entitled *The Past, Present and Future of Mechatronics: A Personal Perspective*.

David Russell Emeritus Professor, Engineering, Penn State University, USA.

He received his B.Eng. degree from the University of Liverpool in 1965 and was awarded his Ph.D. in 1971 by Liverpool Polytechnic (now Liverpool John Moores University) and the University of Manchester for a design study of an intelligent controller for a fast breeder reactor using fluidic control elements.

His teaching at undergraduate and postgraduate levels has spanned over 50 years in the UK and the USA where he was on the Computer Science Faculties at Howard University and Villanova University prior to joining Penn State College of Engineering where he was based throughout at the Great Valley Graduate Campus and from where he recently retired with Emeritus Professor status.

While at Penn State, he implemented several postgraduate degree programmes, including a Master of Systems Engineering, a Master of Software Engineering and a minor in bioinformatics among several others.

He has served as Technical Consultant and Executive for several systems integration corporations and has designed and implemented mechatronic factory information systems throughout North America. He is Author of the Springer book *The BOXES Methodology* which is based on his long-term research interest in nonlinear black box control.

He lectures worldwide and is Author of over 140 technical papers and presentations. He is Chartered Engineer and Fellow of the Institution of Mechanical Engineers (IMechE), past Fellow of the Institution of Electrical Engineers (now the Institution of Engineering & Technology) and Fellow of the British Computer Society (BCS). He is also Long-Time Regional Editor for the Americas for the Springer *Journal of Advanced Manufacturing Technology.*

He is International Member of the Mechatronics, Information and Control Group (MICG) and Mechatronics Forum within the IMechE.

Philip Moore Visiting Professor, University of Skövde, Sweden.

He studied Production Engineering and Management at Loughborough University whilst working in the automotive industry for GKN. He then studied for a Ph.D. in robotics at Loughborough University, before joining as an academic member of staff, when he was Founder Member of the Manufacturing Systems Integration (MSI) Research Institute.

He spent 17 years at De Montfort University (DMU) where he was Director of Research, Professor of Mechatronics, Director of the Mechatronics Research Centre and Founder of the Intelligent Machines and Automation Systems (IMAS) laboratory. He has held his post as Visiting Professor of intelligent automation at the University of Skövde since 1994, where he jointly started the Centre for Intelligent Automation, subsequently the Virtual Systems Research Group.

He was Pro-Vice Chancellor (Research & Innovation) at Falmouth University between 2012 and 2017, where he led the management of all research and development activities, and was also Chair of the Centre for Smart Design, before retiring in 2017. He is now on the Board of Directors and Vice-Chair of Cornwall Mobility a registered charity, also acts as Consultant for Smart Connected Living (SCL) Ltd. and was on the Board of Directors of the Smart Home and Building Association (SH&BA) for over ten years until 2016.

His own research focuses on digital manufacturing and automation, smart digital technologies and their integration with sustainable design where he has won tens of millions in research funding from UK and EU agencies including: EPSRC, ESRC, BBSRC, Innovate UK, EU FP, British Council, Royal Society, Regional Development Agencies and industry. He has supervised some 50 Ph.D. programmes to completion and has been external examiner on well over 50 occasions.

He is Vice-Chair of the Mechatronics, Informatics and Control Group at the Institution of Mechanical Engineers and chair of the Mechatronics Forum.

Reinventing Mechatronics

Xiu-Tian Yan and David Bradley

Abstract Mechatronics was initially conceived as a design based engineering concept which related to the impact of computing processing power and electronics on the design and operation of a wide range of mechanical systems. Since then, there have been dramatic changes in system functionality and capability, driven largely by developments in computing and information technologies, resulting in systems of increasing capability and complexity. More recently, the growth of system concepts such as that of Cyber-Physical Systems and the development of the Internet of Things and The Cloud have also impacted on the nature and role of mechatronics, creating opportunities for step-change improvement, particularly in respect of the provision of smart components and sub-systems and their configuration. These have brought with them new challenges in areas such as system modelling, user privacy and machine ethics, all of which relate to the need, and opportunity, for mechatronics to reinvent itself both contextually and conceptually while retaining its core concepts of system integration. The current chapter therefore brings together these and other issues to provide the framework for the subsequent chapters.

Introduction and Background

The 50 years since mechatronics was initially conceived as a design based engineering concept which expressed the increasing capability and capacity for computing power and electronics to impact on the design of what were still essentially mechanical systems have seen dramatic changes in system functionality and capability, driven largely by developments in computing and information technology. The extent of these changes within the context of the core mechatronics technologies is illustrated by Fig. 1 which suggests, albeit in a somewhat arbitrary fashion, the different levels of technical evolution that have taken place, taking 1970 as the reference base.

X.-T. Yan
University of Strathclyde, Glasgow, UK

D. Bradley (✉)
Abertay University, Dundee, UK
e-mail: dabonipad@gmail.com

X.-T. Yan et al. (eds.), *Reinventing Mechatronics*,
https://doi.org/10.1007/978-3-030-29131-0_1

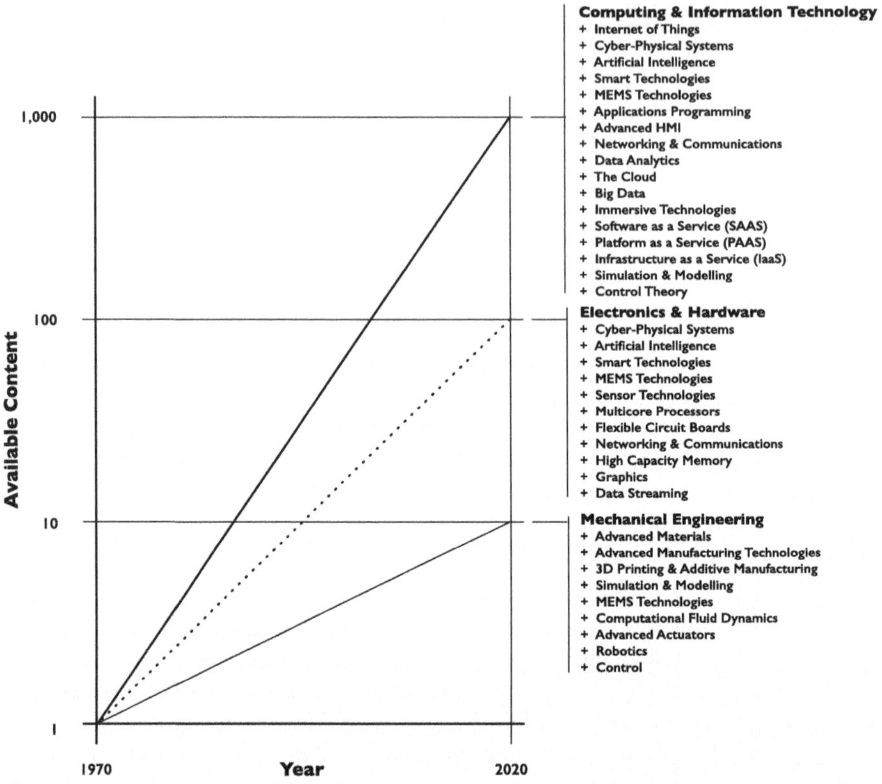

Fig. 1 Growth in available technical content

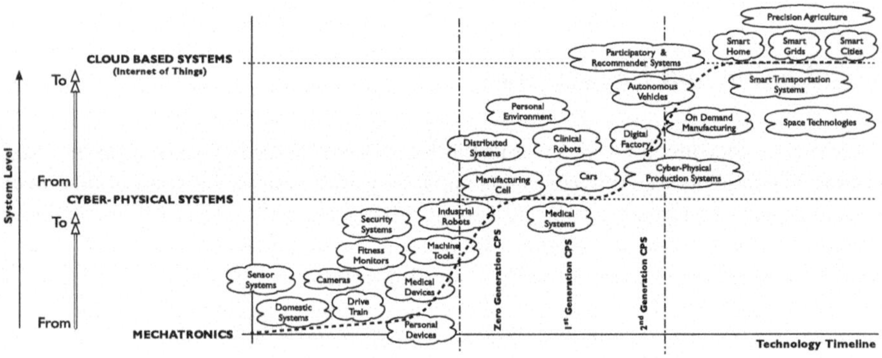

Fig. 2 From mechatronics to cyber-physical systems and the internet of things. Modified & adapted from Hehenberger et al. [3]

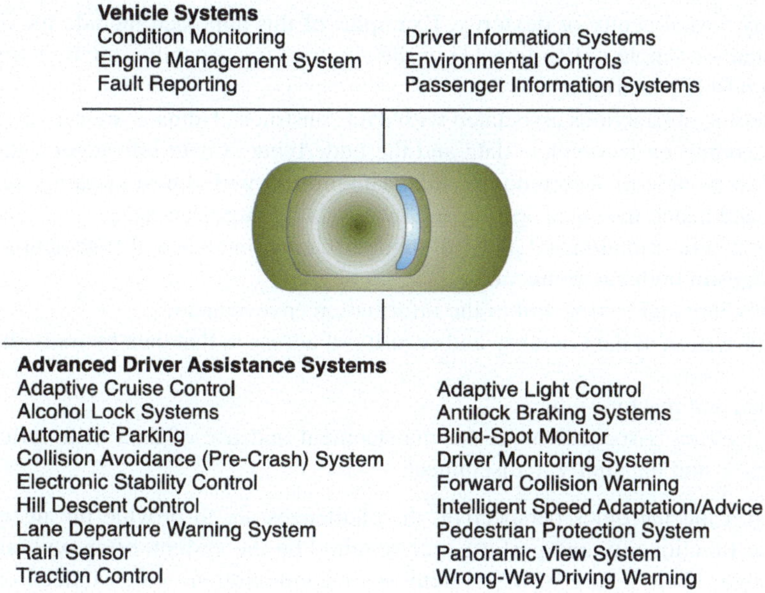

Fig. 3 Vehicle systems

Figure 2 than suggests how these technologies have come together in enabling the implementation of increasingly complex systems, each of which however have mechatronic components and sub-systems at their foundation. Specifically, these mechatronic elements are increasingly configured as smart objects providing access to information as well as being the primary sources of real-world actuation.

Systems such as those suggested at the Cyber-Physical Systems and Cloud levels of Fig. 2 are structured around the large scale interconnection of individual mechatronic smart objects associated with a range of physical parameters. The information gathered can then be used to:

- Monitor the behaviour of objects in both space and time.
- Monitor and control interactions within the information and physical domains.
- Facilitate advanced data analysis and visualisation to support system operation and user interaction.

A number of issues can now be identified as being associated with the design of such systems including:

- Managing complexity and the ability to communicate this between members of the design team. Here, modelling methods have significant potential to act as a transactional medium between team members, enabling concepts expressed in one technical domain to be related to decisions taken in other domains.
- Interface design and operation, including the design and development of smart or adaptive interfaces which '*learn*' and adapt to individual preferences and requirements. This is also associated with the requirement to accommodate different levels

of, say, visual acuity or dexterity. Examples of this approach include the *Siri*™ assistant on Apple's *iPhone*™ [1] or the eye-tracking capability on the Samsung *Galaxy*™ [2] smart phone.
- Modelling abstractions associated with data constructs. Humans are very efficient at reasoning on incomplete data and the underlying system information models need to be able to accommodate this. This in turn will almost certainly require new, and faster, means of sorting and investigating large data sets.
- Integrated information and physical models to coordinate in real-time actions and activities in both environments.
- Prototyping and testing within the information environment.
- The provision of data security and security of access to that data by individuals.
- Privacy issues.
- Human and machine ethics.
- Social issues associated with the management and use of data in a variety of contexts, and the ability to disconnect.

From a mechatronics perspective, the challenges are to provide the necessary levels of functionality and performance required by the system, often without any knowledge of the nature of the system or its configuration, which indeed can be changed autonomously according to both context and need.

Reinventing Mechatronics

Mechatronics has consistently evolved and developed over its 50 year life, adapting in particular to the changes in processing power which facilitated systems of increasing sophistication. The last 10 years have however seen a further transition in the addition of integrating levels of AI to systems which would previously have been considered mechatronic in nature, and the associated ability of such systems to communicate with other systems through the medium of The Cloud. To place this into context, consider the vehicle systems of Fig. 3 in which the majority are essentially mechatronic in nature [4, 5]. However, such systems may well also be operating in a wider context involving communication between vehicles to ensure operational safety and efficiency. Consider therefore the situation of Fig. 4 in which two vehicles are approaching a blind junction with no priorities. By allowing the vehicles to communicate, and hence to negotiate, they can both reach and cross the junction with minimum delay and no risk of collision.

Similar instances of integrating what are still fundamentally mechatronic devices, components, systems and sub-systems within the wider context of Cyber-Physical Systems and The Cloud can be found in many environments including agriculture and precision farming, manufacturing technologies, robotics technologies, medical devices such as prostheses and space based systems, all of which, along with vehicle systems, are subjects of chapters in this book.

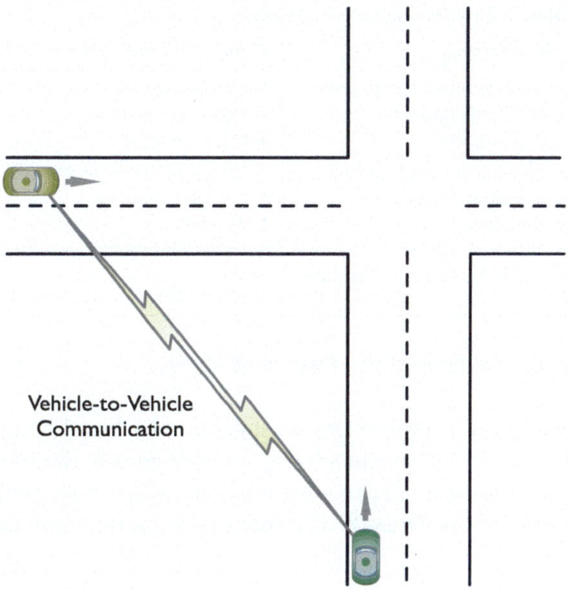

Fig. 4 Negotiation between vehicles

With mechatronics now providing key components for these larger systems, there is a need for it to rethink, and hence to reinvent itself, within this wider framework, almost in the form of service provision, providing the intelligent and smart building blocks for more complex systems. This does not mean that the fundamental technological basis of mechatronics itself must, or needs to, change in any significant way other than recognising the balance of technologies. Rather what is needed is for the mechatronic designer to be able to accommodate a wider range of applications at the fundamental mechatronics level.

Challenges

Privacy and Security

Mechatronic systems and sub-systems are likely to be involved in the collection and distribution of a wide range of data, including on the behaviour of the system user. This data will then be used by other parts of the system to determine its response and behaviour and may well also be integrated with data from other users for analysis and the extraction of new knowledge. In doing so, it raises concerns over user privacy and system security [6, 7] which will need to be addressed both by the designer of the mechatronic elements of any system, as well as the system integrators. Table 1 provides an outline of some of the privacy issues that might be of concern at different system levels.

Table 1 Privacy and security issues

Mechatronics	Cyber-Physical Systems	The Cloud
• System access & security • Data collection & transmission • Monitoring & recording • Location & tracking • Reporting • Personal devices • Data validation & checking	• Device integration • Data aggregation • Data analysis • Data transmission security • Decision making • Models • Data validation & checking	• Data storage, back-up & security of access • Big Data analysis • Big Data algorithms • Data aggregation • Recommender systems • Actions • Data validation & checking

Industry 4.0 and the Future of Work

Industry 4.0 (I4.0) refers, as suggested by Fig. 5, to the future integration and computerisation of manufacturing technologies in the form of smart factories in which systems communicate and cooperate in real-time with each other and with human users and operators [8–13]. Industry 4.0 is thus structured around the following core concepts:

- *Connectivity*—The interconnection and communication of multiple devices and systems as well as people through the medium of the Internet of Things and related technologies. Data is collected in real-time from across the entirety of the manufacturing process to facilitate operation.
- *Transparency*—The transparency of information associated with Industry 4.0 supports decision making at all levels within an organisation.

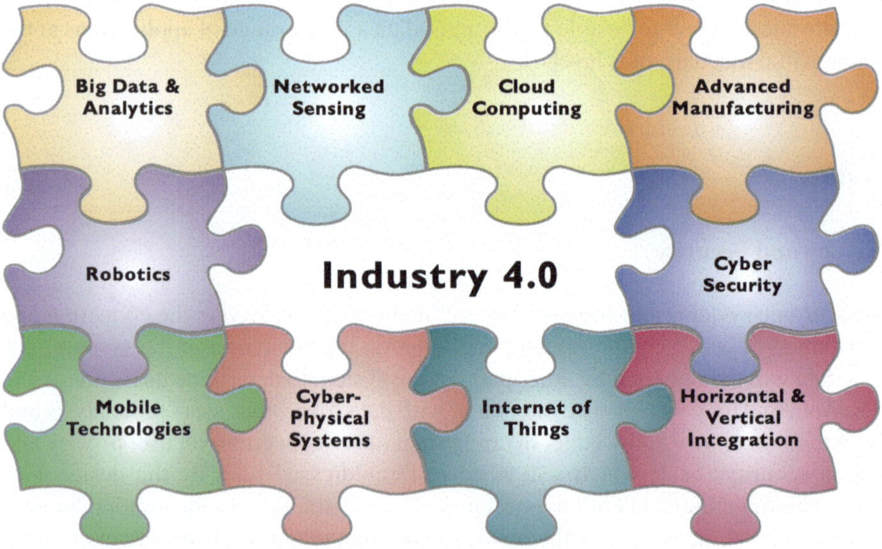

Fig. 5 Core elements of industry 4.0

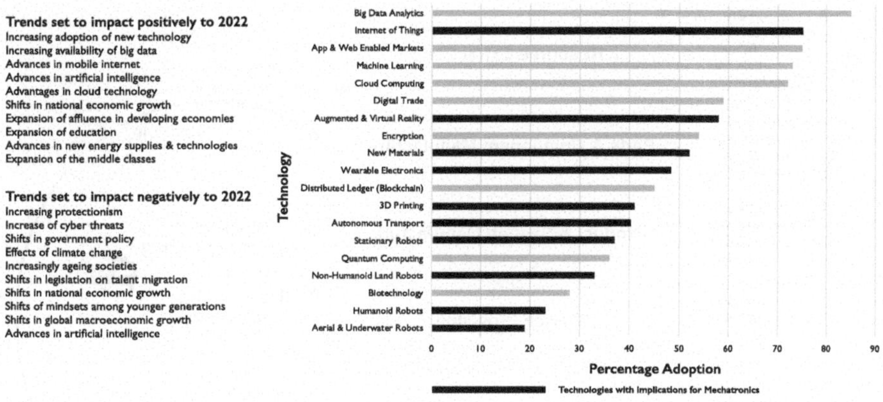

Fig. 6 The future of work. Adapted from [17]

- *Support*—Integrated support systems enable the aggregation and visualisation of information to support decision making and problem solving as well as reducing the burden on individual operators.
- *Decision Making*—AI based systems with the capability of local decision making support enhanced levels of automation.

This however brings with it a number of challenges ranging from increased levels of human-robot collaboration to the changing nature of work associated with Industry 4.0 [14–17]. In this context, the World Economic Forum [16] has suggested both trends set to have positive and negative impacts on the nature of work and key technology drivers, many of which, as is seen in Fig. 6, have implications for the way in which mechatronic systems are perceived, conceived and designed, as well as for the education and training of mechatronic engineers.

Sustainability and the Circular Economy

Issues of sustainability and the impact on society of models such as that of the circular economy of Fig. 7 are becoming of increasing importance, and will impact on the way in which systems, and particularly mechatronic systems, are designed and used [18–22]. This is especially the case with respect to domestic systems where again mechatronics will play a significant role, as for instance in the integration of built-in-self-test facilities within domestic appliances to support a move to maintenance at need rather than by time [23–27]. This shift is in turn likely to be associated with a shift in design approaches to emphasis and facilitate repair and maintenance rather than the simplification of the manufacturing and assembly processes.

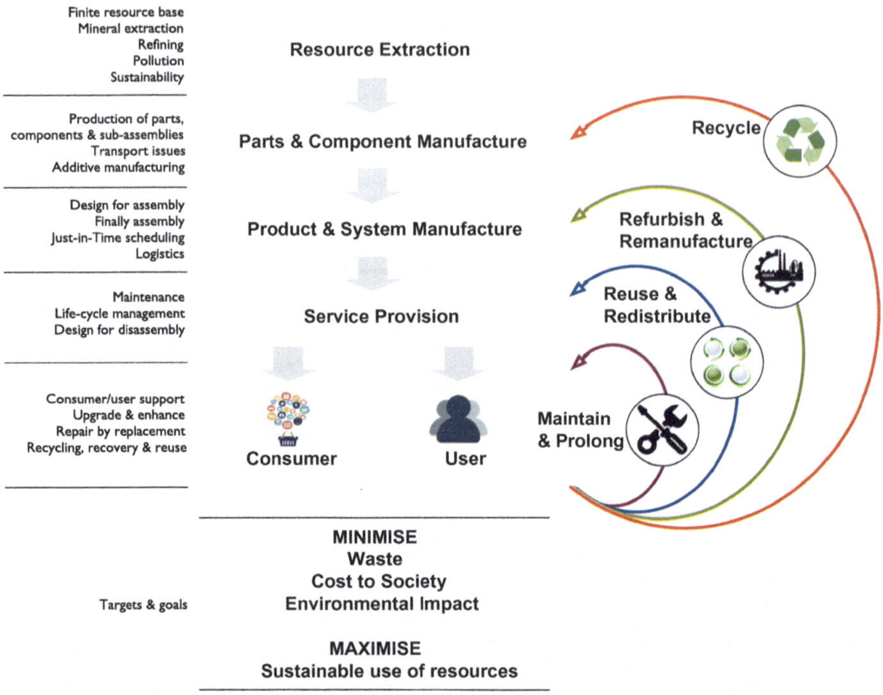

Fig. 7 The circular economy

Education

Changes in society such as those suggested above and driven by developments in technology will place additional demands on the educational process for engineers, and for mechatronic engineers in particular given the integrated and integrating nature of their discipline [28, 29]. This will require a greater awareness of the systems issues associated with Cyber-Physical Systems and those related to operation within a Cloud-based environment, and including increased emphasis on user issues such as security and privacy. All without losing sight of the need to provide a robust and comprehensive foundation in the core disciplines.

Summary

Over the last 50 years the concept of mechatronics as an engineering discipline has gone through a process of continuous change and evolution to adapt to changes in technology. With the evolution of Cyber-Physical Systems and Cloud-based systems, the drivers for change have become more imperative and pressing, even though the underlying technical basis for mechatronics remains to a very significant degree the same, while acknowledging an inevitable shift in emphasis.

The chapters that make up the body of the book were selected from the papers presented at the Mechatronics Forum *Mechatronics 2018* Conference held at the University of Strathclyde in September 2018 to illustrate the dynamic nature of the subject area, and to illustrate this with research based examples, but also to emphasis the underlying basis of mechatronics as an engineering subject discipline.

References

1. Hoy, M. B. (2018). Alexa, siri, cortana, and more: An introduction to voice assistants. *Medical Reference Services Quarterly, 37*(1), 81–88.
2. Sinha, A., Das, R., Saha, S. K. (2018). Eyegaze tracking. Handheld Devices. In: *5th IEEE International Conference on Emerging Applications of Information Technology (EAIT)* (pp. 1–5).
3. Hehenberger, P., Vogel-Heuser, B., Bradley, D. A., Eynard, B., Tomiyama, T., & Achiche, S. (2016). Design, modelling, simulation and integration of cyber physical systems: Methods and applications. *Computers in Industry, 82,* 273–289.
4. Bengler, K., Dietmayer, K., Farber, B., Maurer, M., Stiller, C., & Winner, H. (2014). Three decades of driver assistance systems: Review and future perspectives. *IEEE Intelligent Transportation Systems Magazine, 6*(4), 6–22.
5. Cades, D. M., Crump, C., Lester, B. D., Young, D. (2017). Driver distraction and advanced vehicle assistive systems (ADAS): Investigating effects on driver behaviour. *Advances in Human Aspects of Transportation* (pp. 1015–1022). Springer.
6. Zhang, D. (2018). Big data security and privacy protection. *Advances in Computer Science Research, 77,* 275–278.
7. Jain, P., Gyanchandani, M., & Khare, N. (2016). Big data privacy: A technological perspective and review. *Journal of Big Data, 3*(1), 25.
8. Hermann, M., Pentek, T., Otto, B. (2016). Design principles for industrie 4.0 scenarios. In: *49th IEEE Hawaii International Conference on System Sciences (HICSS)* (pp. 3928–3937).
9. Lom, M., Pribyl, O., Svitek, M. (2016). Industry 4.0 as a part of smart cities. *In: 2016 Prague Smart Cities Symposium* (pp. 1–6).
10. Mrugalska, B., & Wyrwicka, M. K. (2017). Towards lean production in industry 4.0. *Procedia Engineering, 182,* 466–473.
11. Schuh, G., Anderl, R., Gausemeier, J., ten Hompel, M., Wahlster, W. (2017). *Industrie 4.0 maturity index: Managing the digital transformation of companies.* Utz Verlag GmbH. ISBN-13: 978-3831646135.
12. Schwab, K. (2017). *The fourth industrial revolution.* Portfolio Penguin. ISBN-13: 978-0241300756.
13. Gilchrist, A. (2016). *Industry 4.0: The industrial Internet of Things.* A press. ISBN-13: 978-1484220467.
14. Frey, C. B., & Osborne, M. A. (2017). The future of employment: How susceptible are jobs to computerisation? *Technological Forecasting and Social Change, 114,* 254–280.
15. Peters, M. A. (2017). Technological unemployment: Educating for the fourth industrial revolution. *Educational Philosophy and Theory, 49*(1), 1–6.
16. World Economic Forum. (2018). Eight Futures of Work: Scenarios and their Implications, White Paper at www3.weforum.org/docs/WEF_FOW_Eight_Futures.pdf. Accessed March 7, 2019.
17. World Economic Forum. (2018). The Future of Jobs Report 2018 at www.weforum.org/reports/the-future-of-jobs-report-2018. Accessed March 7, 2019.

18. Den Hollander, M. C., Bakker, C. A., & Hultink, E. J. (2017). Product design in a circular economy: Development of a typology of key concepts and terms. *Journal of Industrial Ecology, 21*(3), 517–525.
19. De los Rios, I. C., & Charnley, F. J. (2017). Skills and capabilities for a sustainable and circular economy: The changing role of design. *Journal of Cleaner Production, 160,* 109–122.
20. Homburg, C., Schwemmle, M., & Kuehnl, C. (2015). New product design: Concept, measurement, and consequences. *Journal of Marketing, 79*(3), 41–56.
21. MacArthur, E. (2013). Towards the circular economy. *Journal of Industrial Ecology, 2,* 23–44.
22. Preston, F. (2012). *A global redesign? Shaping the circular economy*. London: Chatham House.
23. Cooper, T. (Ed.). (2010). Longer Lasting Products—Alternatives to the Throwaway Society. Gower. ISBN-13: 978-0-566-08808-7.
24. Gaver, B., Dunne, T., & Pacenti, E. (1999). Design: Cultural probes. *Interactions, 6*(1), 21–29.
25. Hassenzahl, M. (2013). User experience and experience design. *Encyclopedia of Human–Computer Interaction*, 2nd ed.
26. Spangenberg, J. H. (2013). Design for sustainability (DfS): Interface of sustainable production and consumption. *Handbook of sustainable engineering*, 575–595.
27. Torry-Smith, J. M., Qamar, A., Achiche, S., Wikander, J., Mortensen, N. H., & During, C. (2013). Challenges in designing mechatronic systems. *Journal of Mechanical Design, 135*(1), 11.
28. Johansson-Sköldberg, U., Woodilla, J., & Çetinkaya, M. (2013). Design thinking: past, present and possible futures. *Creativity & Innovation Management, 22*(2), 121–146.
29. Lammi, M., & Becker, K. (2013). Engineering design thinking. *Journal of Technology Education, 24*(2), 55–77.

Reinventing Mechatronics—A Personal Perspective

Stephen Sanders

Abstract Robotic systems were first used at the nuclear fusion research project JET in the early 90's to re-configure and maintain the many complex and delicate in-vessel components. The complex nature of the in-vessel environment and the many remote maintenance tasks required the development of highly flexible robotic systems that relied on having a person 'in-the-loop' together with operating methodologies that considered all aspects of the plant and maintenance task. This approach has now been successfully used for the development of remote handling (robotic) systems at both Dounreay in the development of the 'Shaft Intervention Platform' and for fuel debris retrieval at the stricken Fukushima Daiichi nuclear reactors.

Background

Stephen was a founding member of Oxford Technologies Ltd (OTL) in 2000 after having worked in the Remote Handling group at the Joint European Torus (JET) project. Initially involved in the installation of over 100 diagnostic systems inside the JET machine, once it became active, he joined the remote handling group where he was involved in several extensive remote handling campaigns.

Since 2000, as a remote handling expert, he has helped deliver major remote handling contracts in Nuclear Decommissioning, Nuclear Fusion and High Energy Physics. In 2010 came a move to marketing and business development for OTL and in 2013 became Director of marketing & business development.

Post-sale of OTL to Kurion Inc. in December 2015 and Kurion's subsequent sale to Veolia in 2016, he took the position of Global Access Marketing & Business Development and today leads Remote Handling for Veolia Nuclear Solutions as Strategic Director for Remote Operations.

S. Sanders (✉)
Veolia Nuclear Solutions (UK), Didcot, UK
e-mail: stephen.sanders1@veolia.com

X.-T. Yan et al. (eds.), *Reinventing Mechatronics*,
https://doi.org/10.1007/978-3-030-29131-0_2

Joint European Torus

My career in remote handling and robotic systems began in the mid 90s when I joined the Joint European Torus (JET) project at Culham, Oxfordshire, UK after spending the previous 15 years working on the development of computer systems. JET was, and is still today, a world leading research project into nuclear fusion, the search for an ultimate source of clean energy for humanity by fusing two isotopes of hydrogen, deuterium and tritium and in the process releasing helium and high energy neutrons. While my early role in the project was the manual installation of over 100 diagnostic systems into the JET machine, working in a full pressure suit. Later when the machine became active and started to generate high energy plasmas, it switched to remote robotic systems for machine maintenance (Fig. 1).

These '*remote handling*' systems comprised a sophisticated mix of mechatronics, control systems, human–machine interfaces and later, virtual reality software. But at JET the real spur to innovation of the remote systems was the internal complexity of the JET machine environment, a 6 m radius by 3 m high torus covered with many thousands of delicate components and graphite tiles built to resist the 100 million degree plasmas that were generated and held inside a 3.45 T magnetic bottle whizzing around the torus at high speed. To allow for the massive thermal gradient the machine was built from 8 separate octants joined by steel bellows to allow for expansion and contraction during and after machine operation. After each 'scientific' operational period, usually lasting 2 years, the machine was shut down and re-configured. Its construction in 'octants with expanding bellows' meant the physical form changed

Fig. 1 JET machine with robotic boom carrying slave tele-manipulator on the left of picture. Image courtesy of Eurofusion

Fig. 2 JET octant 1 robotic boom carrying a '*task module*' via remote docking interface (left) and mascot slave tele-manipulator on octant 5 robotic boom (right). Images courtesy of Oxford Technologies & Eurofusion

constantly during its life and so was not a suitable environment for rigid robotic systems requiring high degrees of positional accuracy. We required flexibility in navigating the internal torus and human-levels of dexterity for many of the remote maintenance tasks on systems designed to be maintained by human hands. We also required the ability to adapt the remote systems to the ever changing 'un-constrained' internal environments of the JET machine.

The result was an 18 degree of freedom, 12 m long cantilevered robotic boom system with an end-effector cable of carrying loads up to 500 kg and positioning them with millimetre accuracy or carrying a second robotic bi-lateral telemanipulator offering a further 7 degrees of freedom on each of its two arms, capable of carrying 20 kg each or 100 kg using a chest mounted hoist (Fig. 2).

The system was operated several hundred metres away in a control room through specially developed human–machine interfaces and in the case of the tele-manipulator, by an operator sat at the master manipulator using screens to view the slave mounted onto the robotic boom.

Commissioning the system involved a 1000 h continuous operational period to reveal any infant mortalities in the system. The JET remote handling boom has now operated for over 20 years with only one significant failure of carriage bearings.

A key methodology in the development of such remote handling systems and its operational remote tasks is the constant validation of equipment and tools using digital simulations, bench trials and eventually full scale mock-up trials prior to release for active operations. This process has been adopted on all Oxford Technologies remote handling system developments.

Of course, not all remote handling systems require such complexity and one should resist the urge to add complexity where not needed, despite the intellectual challenge. Clear task requirements are essential. Again, OTL developed a methodology for defining a clear set of requirements on which the system specification was built. This methodology involved three stages, to understand and define the environment and plant on which the remote task is based, to define the task itself and finally perform a remote handling compatibility assessment on the task. As an example, one could consider a car maintenance task where you cannot physically reach the components requiring maintenance, so we first clearly define the environment and plant (car) characteristics. These could be that the car is inside a poorly lit garage, with the bonnet open and a drawing or model of the engine is available. The temperature and condition of the engine may be relevant. Secondly, we define the nature of the maintenance task, perhaps replacing an engine sensor buried deep in the engine compartment. We require information on the sensor, how its retained and degrees of freedom to remove it. Now we can consider the task for its compatibility with the available remote handling systems and highlight any special equipment or tooling that has to be developed. We will have to take into account the working environment, availability of services and viewing/lighting needs. We may now create a clear specification of the remote task and system requirements to allow concepts to be developed. We will also have to develop a methodology by which the remote system can be deployed to complete the task. All concepts can be tested through the use of digital simulations that allow ideas for new tooling to be quickly trialled or kinematics of the remote system and tooling to be evaluated. Such discipline ensures all factors are considered when creating a specification on which the equipment concepts are based.

Dounreay

These methodologies were used to good affect when developing a design scheme for the decommissioning of the Dounreay shaft. Built in 1956 the 64 m deep, 4.5 m diameter shaft with a tunnel at its base running out to sea was originally created to flush active effluent out to sea, but sense prevailed and it was instead used for the storage of intermediate level radioactive waste when the stub tunnel at its base was capped at 20 m.

Due to the long history of the Dounreay shaft and poor record keeping in the early years, the contents, lower down in particular, were poorly defined but included a variety of canisters, metal tanks, cables and large lumps of concrete resulting from an explosion in the shaft due to a build-up of gas in the 70's. So one might consider this the most challenging of environments into which we would have to deploy robotic remote handling systems and certainly it is seen as one of the most challenging decommissioning projects in Europe. Nevertheless, the same methodologies were applied to clearly define the environment of the shaft, its construction method, any headworks, the nature of the hazards including radioactivity, risk of explosion, water (the shaft was kept flooded) and of course to catalogue the contents, though most were believed to be heavily degraded by time and water. The task list was based not just on how to retrieve the identified contents but also interfaces with associated headworks equipment such as cranes, waste sorting systems, maintenance systems and how to clean out the 20 m stub tunnel.

A number of concepts were developed and evaluated using digital mock-ups of the shaft and typical contents. Of particular concern was the kinematic form of any robotic arms used and how best to retrieve the heavy concrete blocks and any large,

Fig. 3 Dounreay 65 m deep by 4.5 m diameter shaft with 20 m stub tunnel (left) and the shaft intervention platform to be deployed over the shaft for remote waste retrieval (right). Images courtesy of DSRL & Oxford Technologies

heavy items discovered in the shaft. The final optimised concept involved the use of a horseshoe shaped platform that could operate in conjunction with a crane mounted petal grab. Slung underneath the horseshoe platform were two telescopic robotic arms with mobile wrists onto which were remote docking interfaces for a wide range of tooling. Bench trials were conducted using a full scale prototype arm at OTL's assembly and test facility which incorporates a 10 m × 5 m × 4.5 m deep pit as well as trials to clean out sludge from the stub tunnel using a high pressure hose deployed by the robotic arms (Fig. 3).

The complete '*Shaft Intervention Platform*' together with its deployment system has now been built and is due to undergo factory acceptance trials in 2019.

Fukushima Daiichi

Today, the company is focussed on solving the remote access challenges posed by the heavily damaged nuclear reactors at Fukushima Daiichi in support of Japanese government agencies and industry. The prime tasks are to investigate the interior condition of the reactors, locate the fuel which is thought to have melted through the reactor core and eventually to retrieve the fuel debris. Remote access to the chosen reactor, Unit 2, is via an horizontal port of 600 mm diameter. The reactor is approximately 20 m diameter and so any remote intervention system has to travel around 16 m to reach the core and be able to deploy sensors and tooling a further 8 m depth to locate melted fuel. A further requirement was to inspect the circumference of the internal pressure containment vessel and so a total reach for the remote system was determined to be 22 m.

The development of robotic systems that can navigate such an unconstrained environment with high levels of radiation, humidity and debris, places considerable challenges on the design team and their choices of suitable technologies and materials. These must be capable of not just surviving such an extreme environment but of

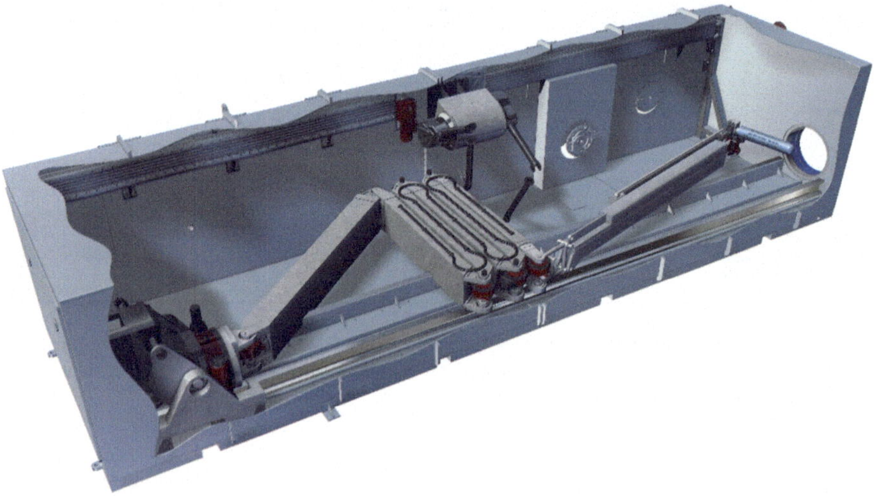

Fig. 4 MHI investigation boom system incorporating 21.9 m long, 18 degree-of-freedom robotic boom inside an air tight enclosure with dexter slave manipulator for remote handling of tools and sensors (oxford technologies—veolia nuclear solutions (UK))

delivering high levels of positional accuracy and repeatability required by the sensors while delivering a payload at 22 m reach through a 600 mm port.

The remote handling robotic boom technologies previously developed for maintenance of nuclear fusion machines by OTL formed the basis of the remote systems currently in development for Fukushima Daiichi but the environment and task challenges required the further development of such boom technology, particularly the requirement to operate over several metres in the vertical plane as well as to navigate the 20 m reactor containment vessel. Over a thousand man hours has been spent in development of the boom link actuators alone ensuring they deliver the required positional performance and are sealed against the environmental contaminants (Fig. 4).

The system is currently in manufacture with operational trials expected mid 2019 inside a specially constructed full scale mock-up of the damaged reactor.

Comment

While such projects test engineering teams to the limit they are also the lifeblood on which new and exciting mechatronic developments are based, they are what, as engineers, we live and breathe for.

From Mechatronics to the Cloud

David Bradley, David Russell, Peter Hehenberger, Jorge Azorin-Lopez, Steve Watt and Christopher Milne

Abstract At its conception mechatronics was viewed purely in terms of the ability to integrate the technologies of mechanical and electrical engineering with computer science to transfer functionality, and hence complexity, from the mechanical domain to the software domain. However, as technologies, and in particular computing technologies, have evolved so the nature of mechatronics has changed from being purely associated with essentially stand-alone systems such as robots to providing the smart objects and systems which are the building blocks for Cyber-Physical Systems, and hence for Internet of Things and Cloud-based systems. With the possible advent of a 4th Industrial Revolution structured around these systems level concepts, mechatronics must again adapt its world view, if not its underlying technologies, to meet this new challenge.

How Did We Get to Here?

The concept of a central repository of computational power and associated storage is not new, and systems of this nature were historically used by companies and universities to support remote users [1]. As the internet became increasingly available, so distributed computing power in the form of client-server architectures emerged when information could be streamed to and from a user and the interaction between

D. Bradley (✉)
Abertay University, Dundee, UK
e-mail: dabonipad@gmail.com

D. Russell
Penn State Great Valley, Malvern, USA

P. Hehenberger
University of Applied Sciences Upper Austria, Wels, Austria

J. Azorin-Lopez
University of Alicante, Alicante, Spain

S. Watt · C. Milne
University of St Andrews, St Andrews, UK

X.-T. Yan et al. (eds.), *Reinventing Mechatronics*,
https://doi.org/10.1007/978-3-030-29131-0_3

the computer centre and a remote user eventually evolved into internet transaction packages [2, 3].

The web-based information industry is now invisibly connected to remote servers, data banks, and platforms with The Cloud providing application and distribution centres [4]. Information-heavy industries such as banks, investment houses, and government agencies were frontrunners in this mode of operation followed by on-line retailers, virtual video game systems, voice-over internet protocols (e.g. Skype®) and search engines such as Siri® and Google®. This was followed by the introduction of remotely accessible smart devices or objects capable of being used for a range and variety of purposes such as environmental control, the opening and closing of garage doors, or activating and querying alarm systems [5].

Mechatronics, Cyber-Physical Systems, the Internet of Things & the Cloud

As suggested by Fig. 1, mechatronics was a major driver of and for the 3rd Industrial Revolution, itself structured around computing, information technology and robotics, that began around 1970. However, the advent of a 4th Industrial Revolution configured around Cyber-Physical Systems (CPS) [6], the Internet of Things (IoT) [7], Cloud technologies [8] and elements of Big Data [9] means that mechatronic devices and systems must now be structured such as to provide the intelligent or smart components and objects with which such a system is built.

To put this emerging and evolving role for mechatronics into context, consider the structures of Fig. 2 and the system exemplars of Table 1 which bring together these various elements and identifies the relationships between the layers in the resulting hierarchy. What is then immediately clear is that there is a requirement for on-demand access to a range of resources including infrastructure, software and platforms with the resulting system a dynamic entity with smart objects and users entering and leaving dependent on both context and need. This then means that both data and information become commodities to be traded by the system on request.

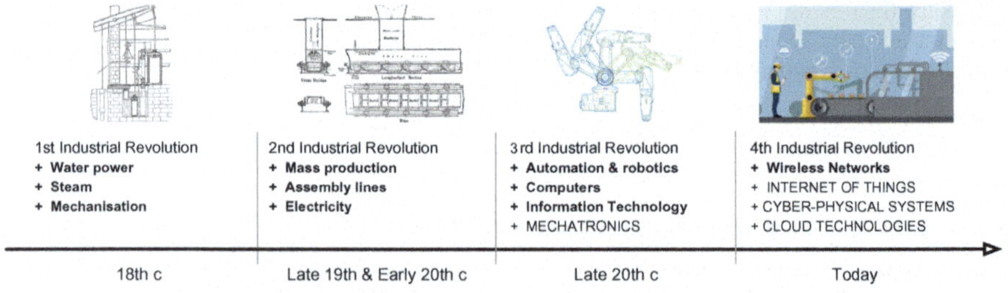

Fig. 1 Timeline of Industrial Revolutions

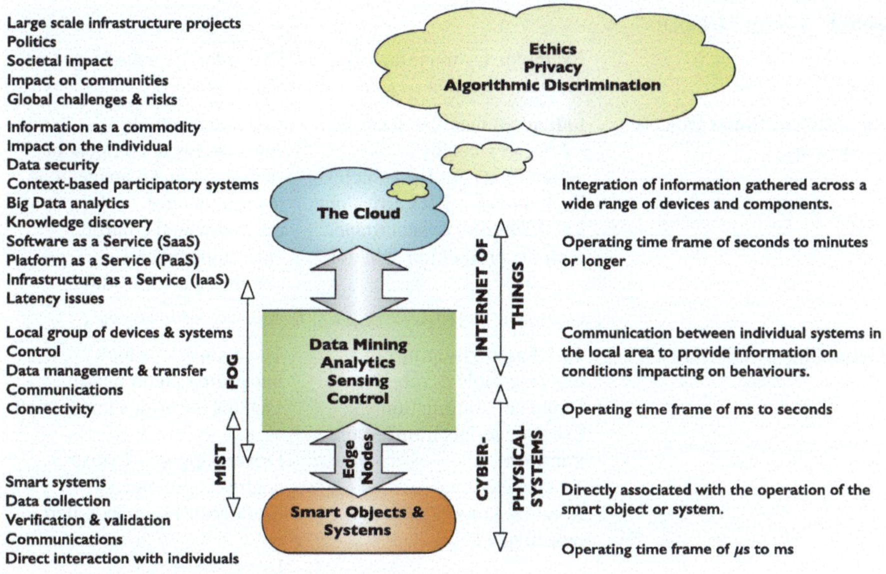

Fig. 2 From mechatronic smart objects and systems to The Cloud

Thus for instance, knowledge that a traffic incident is resulting in delays is only of value to a user if they were intending to travel on affected roads.

Other system elements included in Fig. 2 are:

Edge Nodes—Support the operation of smart devices or components by avoiding the need to send information to the cloud for processing, thus avoiding network latency effects.
Mist Layer—Provides an interface between the local processing associated with the edge nodes and the Fog.
Fog Computing—This is defined by the National Institute of Standards and Technology (NIST) in the US as [10]:

> … a horizontal, physical or virtual resource paradigm that resides between smart end-devices and traditional cloud or data centres. This paradigm supports vertically-isolated, latency-sensitive applications by providing ubiquitous, scalable, layered, federated, and distributed computing, storage, and network connectivity.

Features of Fog Computing include [10–14]:

Distribution—Supports highly distributed services.
Cloud to Things—Fog nodes are positioned close to functional smart objects so that analysis and response times are reduced.
Horizontal Architecture—Supports multiple application domains.
Interoperability—Seamless service support requires the co-operation of different providers implying interoperability and the federation of services.
Mobility—Directly associated with mobile devices.

Table 1 System exemplars

	Exemplar 1—manufacturing system	Exemplar 2—vehicle systems
Mechatronic Smart Objects & Systems	Individual machine tools and robots serviced by autonomous guided vehicles Each tool or robot carries out a task or tasks in accordance with the production schedule	Individual vehicle sub-systems & components such as engine management, traction control, environmental controls, entertainment and communications systems
	Operating time frame of μs to ms and possibly seconds	
Cyber-Physical System	Task based groupings of machine tools & robots Islands of Automation Local materials transfer and transport Task scheduling Condition monitoring & reporting	The complete vehicle integrating all on board systems at various levels such as: Driver support – Cruise control, autonomous headlights, blind-spot alert, etc. Driver assistance – Self-parking, emergency braking, etc. Autonomous operation As determined by installed software
	Operating time frame of ms to seconds and possibly longer	
Internet of Things	Factory wide scheduling Materials transfer & transport Inventory management Materials incoming & despatch	Communication with other vehicles in vicinity Traffic alerts Routing information Parking requests & allocation
	Operating time frame of seconds to minutes and possibly longer	
Cloud	Data storage & analysis Big Data analytics Performance data collection Inventory control Order handling Production planning	Data storage & analysis Big Data analytics Performance data collection Fault detection across fleet Digital twin Traffic management systems
	Operating time frame of seconds to minutes, hours or even days depending on context	

Real-Time Functionality—Analysis in real-time of streamed data.
Sensor Networks—Includes real-time validation and verification.

Design Issues

Conventionally, the design of complex engineering systems followed a structured path in which the design and development stages, including those for sub-assemblies or sub-components, were linked to the implementation stages by validation and verification procedures intended to confirm performance to specification. The net result is that all system elements, hardware, software and firmware, are developed under the overall control and responsibility of the design team. This design pathway is increasingly being challenged by developments in mechatronics and Cyber-Physical Systems (CPSs) and their links to the Internet of Things (IoT) and The Cloud in which the transition from the (mechatronic) component to a CPS and the IoT results in increasing levels of abstraction, limiting the ability of the design team to maintain control over, or even input to, the entirety of the system. This means that those system elements drawn from The Cloud will in general be unknown to the designers of smart (i.e. mechatronic) sub-assemblies and components while still being required to establish and define system functionality [5], a situation illustrated here by Fig. 3 when insertion of a new component can result in an, albeit unintended, system failure.

The range and scope of the challenges facing the designer can be expressed by reference to the World Economic Forum Report '*Deep Shift - Technology Tipping Points and Societal Impact*' [15] which identified the major areas of impact upon society being those set out in Table 2.

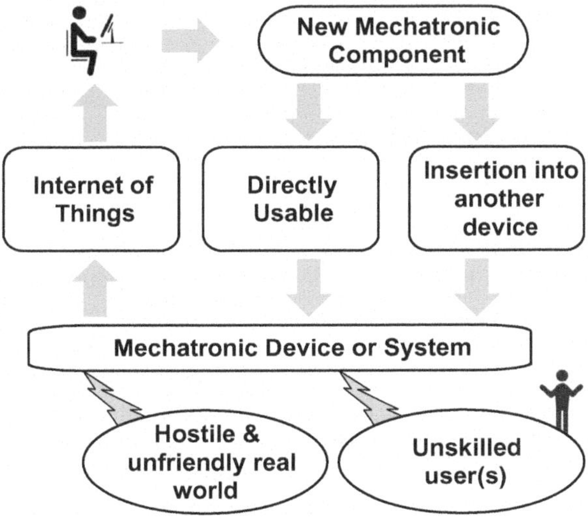

Fig. 3 From design to real-world

Table 2 Potential for societal impact

People & the Internet	How people connect with others, information and the world around them is being transformed. Wearable and implantable technologies will enhance an individual's "*digital presence*", allowing them to interact with objects and each another in new ways
Universal Computing, Communications & Storage	The continued decline in the size and cost of computing and connectivity technologies is driving an exponential growth in the potential to access and leverage the internet. This will lead to the availability of ubiquitous computing power where everyone has access to a supercomputer in their pocket, with nearly unlimited storage capacity
Internet of Things	Smaller, cheaper and smarter sensors are being introduced in homes, clothes and accessories, cities, transport and energy networks, as well as in manufacturing
Artificial. Intelligence & Big Data	Exponential digitisation creates exponentially more data about everything and everyone. The sophistication of the problems that can be addressed, and the ability for software to learn and evolve, is advancing in parallel
Sharing Economy & Distributed Trust	The internet is driving a shift towards networks and platform-based social and economic models, creating not just new efficiencies but also whole new business models and opportunities for social self-organisation
Digitisation of Matter	3D printing as a process that transforms industrial manufacturing and allows for home based production. It also creates a new set of opportunities for human health

What Does What?

The increasingly distributed nature of systems involving the IoT and The Cloud has resulted in consideration of the potential architectures for such systems [17–19] as suggested by Fig. 4. In the context of mechatronics, the associated structural relationships can then be illustrated by Fig. 5 when mechatronics sits at the operational level of machine based systems. This suggests that the mechatronics designer must be primarily concerned with achieving the required functionality and performance at the relevant process layer whilst providing the necessary information and data to feed upwards (in terms of the model) to the CPS, IoT and Cloud levels. The translation and interpretation of data and information as it moves between the process layers is then associated with increasing levels of design abstraction as the nature, structure

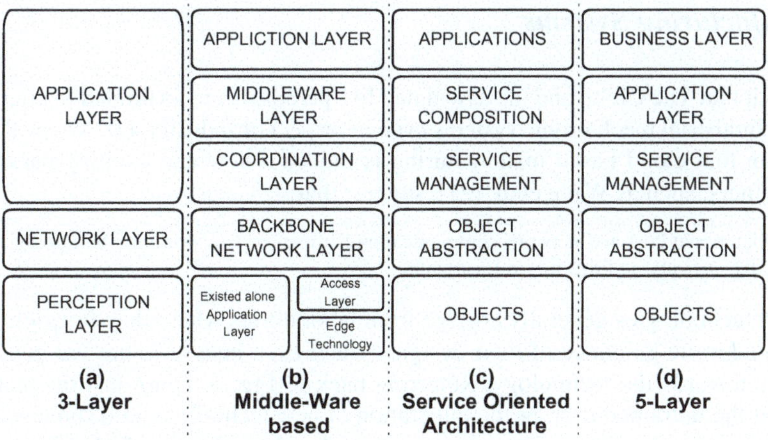

Fig. 4 Potential IoT architectures {after [16]}

Fig. 5 Procedural and process information based system representation

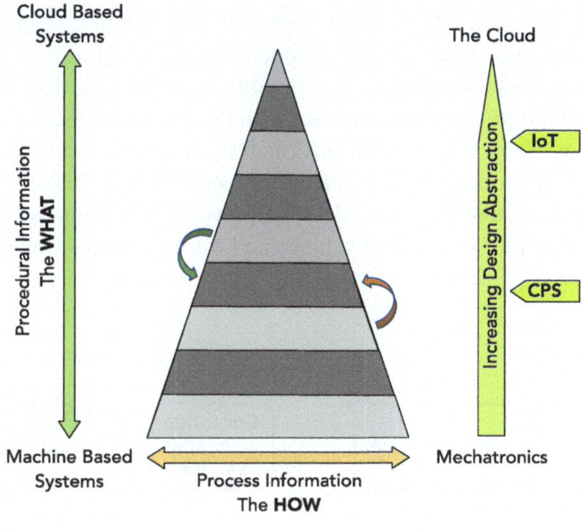

and context of the finally system will generally be unknown at the mechatronics layer(s).

In the context of Figs. 4 and 5, this means that system intelligence is distributed throughout the system from the mechatronics layer(s) to the cloud, with each layer functioning to provide its own specific behavioural contributions within an overall system context. As mechatronic systems are the primary providers of data to the cloud, their design must take account of what is necessary to transmit the data in the form of the associated processes and protocols as well as the levels of translation and interpretation involved.

Manufacturing Systems

Why is it that The Cloud and the associated IoT paradigm are so limited in effectiveness in industrial mechatronic systems even in respect of Industry 4.0? Wu et al. [20] proclaim that cloud based manufacturing is indeed a new and exciting paradigm, while others, such as Wang et al. [21], suggest that:

> Computational cost and network communication (limitations) … present a bottleneck for effective utilization of this new infrastructure.

New technologies generally emerge from laboratories where they function flawlessly under the guidance of their designers, and even beta test sites are generally friendly towards the technology. Referring back to Fig. 3, is not just the learning curve of the untrained user, or the integration of new technology into some existing application, that determines outcomes in service but the influence of what, in contrast to the laboratory, is often a hostile, demanding and unfriendly physical environment.

Since the industrial revolution, automation has increasingly featured in the operation and development plans of manufacturing and smart devices are now found in factories which when operating, promote higher quality products for ever more selective customers. Components such as automated materials handling, packing, autonomous vehicles and flexible groupings of robots and machine tools all contribute to outcomes, yet are usually reliant on local computational support as suggested by Fig. 6a with information access to and by the outside world restricted to order processing, job scheduling and product delivery. Here, local computing supplies schedules, order data, and inventory management with information fed to the shop floor to set up machines and transfer production data.

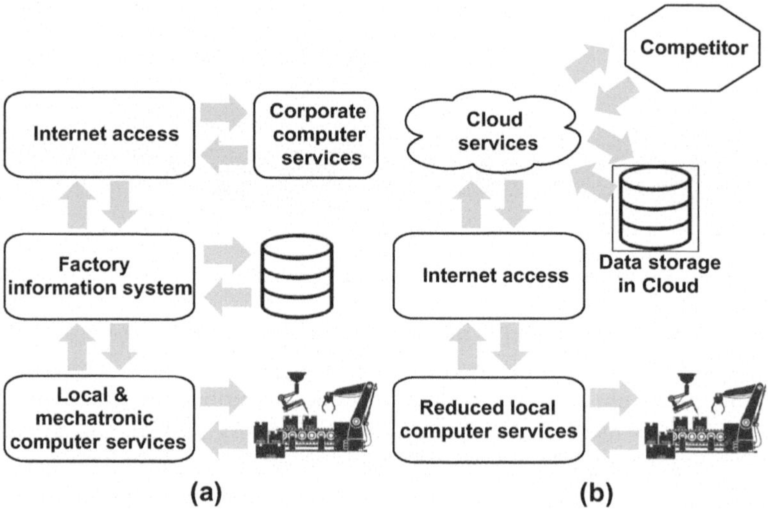

Fig. 6 Manufacturing systems **a** Self contained and **b** Cloud based

In contrast, Fig. 6b illustrates a cloud-based system, and shows the potential risk presented by a competitor being able to access company data. Access to large data sets may have significant commercial value but data centres are very expensive to fund, design and operate and the data they contain is vulnerable to access by others than for whom it was intended. Further, the inherent latency associated with the transmission of data to and from The Cloud means that it cannot, as demonstrated by Figs. 7 and 8, be used to directly control shop floor operations and Cloud-based systems thus tend to be more closely integrated with strategy than operations.

Indeed, the small to medium enterprises (SME) which constitute a very large proportion of manufacturing, may only have one or a two facilities and consequently

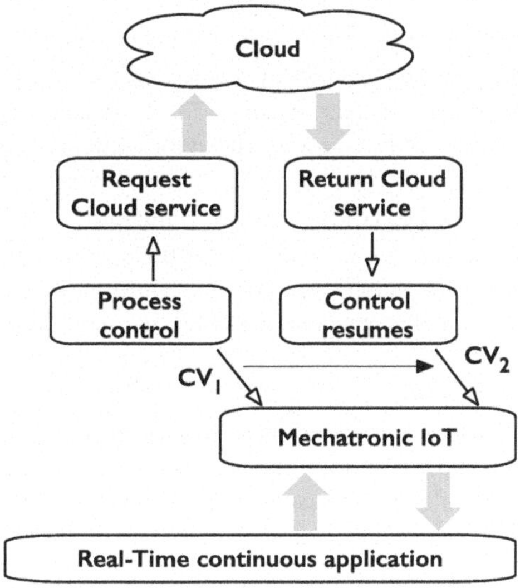

Fig. 7 Cloud-serviced control

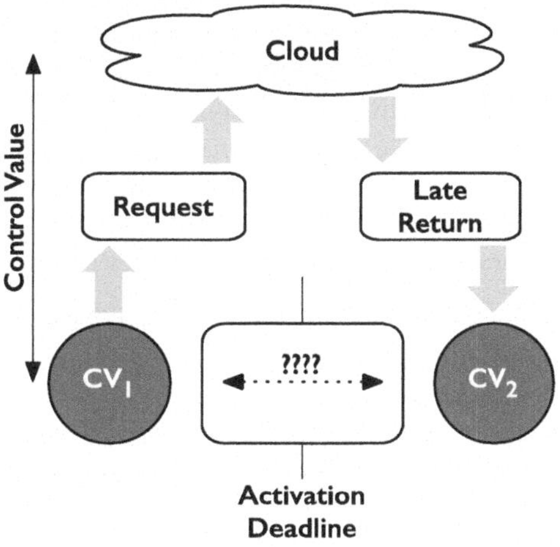

Fig. 8 Timing uncertainty

are not financially or strategically interested in cloud systems and as they are often suppliers of key components to larger clients there is the danger of their secrets being revealed to competitors. The number of such firms is large; in 2014 in the USA, over 97% of manufacturing was associated with firms employing under 500 employees [22], and many have fewer than 50.

Issues & Concerns

Privacy

Historically, less consideration has been given to the '*softer*' or people oriented aspects of individual privacy and there is now a growing imperative to ensure data security and user privacy [23–27] within and as part of the design process [28–30] while ability to apply big data analysis tools to the generated data creates further questions [31].

In respect of the privacy of the individual, as opposed to the security of the system, the '*always on*' society continues to demand greater connectivity at higher speeds. In such an environment, the requirement is that connection to the local network, and hence to the Internet of Things, is essentially seamless. The resulting operational shift from device management to data management suggests that conventional approaches to network security and individual privacy are no longer acceptable or viable, and that securing networks now requires more focus on securing what is important rather than trying to implement a lockdown approach intended to secure everything. This is particularly the case for Cloud-based systems structured around mechatronics, CPS & the IoT which are autonomously communicating and receiving information on behalf of their user.

Many of the devices and systems associated with the IoT have the capability to rapidly accumulate large volumes of personal data, much of which is likely to be held in locations and ways unknown to the user. This data is then subject to the possibility of analysis using the techniques and methods of Big Data [32–34], with a significant risk of impacting on the privacy of individuals [35]. Of concern is the potential to use inference to suggest personal details and behaviour. A simple instance of this is the recommender systems used as a marketing tool by companies such as *Amazon* [36] which use information derived from past customer purchases and search profiles to generate focused advertising. Other examples include:

- The potential to use information derived from, say, traffic routing apps or vehicle systems linked to domestic environmental controls to identify if a house is currently occupied.
- The potential use of information derived from eHealth systems to determine an individual's ability to access or purchase elements of healthcare provision.

The ability to analyse large volumes of data to extract potentially beneficial knowledge, particularly within the context of IoT based applications such as eHealth, for instance to provide an early warning of an impending outbreak of an infectious disease based on consolidated eHealth data, presents a major challenge to the concepts of individual privacy, and hence to system designers. And of course, there is also the potential for other, more nefarious, activities and actions based on accumulated individual data. These concerns have led to the concept of '*Digital or Algorithmic Discrimination*' [5, 37–40] where the use of an individual's personal data within a Big Data algorithm leads to their being in some significant degree being discriminated against, for instance by being denied access to specific services, or being unreasonably targeted in some way. In illustration O'Neill [5] provides (from among others) the following examples:

- For profit colleges in the US used algorithms to generate advertising targeted at poorer and disadvantaged households to enable the college to access government funding at 90%.
- The use of geographically oriented law enforcement management programs such as *PredPol* and *CompStat* led to an emphasis on nuisance crimes in poorer neighbourhoods rather than on more serious crimes elsewhere.

In broad terms, discrimination is defined as the unfair treatment of an individual because of their membership of a particular group and in this context, algorithmic profiling for the allocation of resources can be considered as inherently discriminatory when data subjects are grouped into categories according to selected variables, and decisions made on the basis of subjects falling within defined groups. In this context, machine learning can reinforce existing patterns of discrimination. If these are embodied in the training dataset, then they will be reproduced by the classifier with biased decisions presented as derived from an '*objective*' algorithm. It is a requirement that data controllers act to prevent such discriminatory effects when processing sensitive data which can include or encompass a wide range of personal information as for instance [40]:

- Racial or ethnic information.
- Religious or other beliefs.
- Membership of organisations such as trade-unions.
- Genetic or biometric data.

Whatever the ultimate outcome of the continuing legislative debate over privacy, it is clear that there is an increasing burden on system designers to place privacy at the core of their work, and that this must be reflected in changes to the design process and the associated methods and tools used to support this [41]. This brings with it concerns in relation to the ability of current best practice to accommodate the intent of legislation, and hence to meet guidelines.

The Consumer

The impact of mechatronics is increasingly being felt at the consumer level and mechatronic devices are common in cars (driverless and otherwise), haptic video game consoles, self-focusing automatic cameras, home security systems, smart cities, smart houses, medical robots and a multitude of other applications. Mobile technology enables the locking of house doors remotely and checking the interior remotely while on holiday, or checking how many footsteps a person makes that day as part of a wellness program, and broadcasting that to the web for comparison with others.[1]

The ability to establish an individual's location from a satellite and a mobile phone, be it turned on or not, has introduced many benefits, but also creates a whole new set of privacy issues [42]. For instance, knowledge of an individual's location and behaviour can be used to establish when they are away from home, and hence when their home is vulnerable. The implication is therefore that consumer mechatronics must not only complement the users' world, but must also autonomously adapt to unplanned circumstances and situations while recognising and responding to physical, legal and other constraints to act responsibly within an appropriate time frame.

Ethics

Within the system structure of Fig. 2, ethical issues exist at a number of levels.

Smart Object, Smart Device or Cyber-Physical System

Consider the following scenario involving an autonomous vehicle[2]:

> It's a bright, sunny day and you're sitting back in your self-driving vehicle travelling at the speed limit along a tree lined road. A school bus, driven by a human, is travelling towards you and suddenly veers into your path. There is no time to stop safely, and no time for you to take control of the car.

Does the car:

- Swerve sharply into the roadside trees, possibly killing you but possibly saving the bus and its occupants?
- Perform an evasive manoeuvre around the bus and into the oncoming lane, saving you, but sending the bus into the trees and possibly killing the driver and some of the children on board?
- Hit the bus, possibly killing you as well as the driver and children on the bus?

[1] See for instance www.strava.com (as of October 2018).

[2] Adapted from [43].

Whatever choice is made there is the likelihood that someone will be killed, so what should the autonomous vehicle do? Whatever decision is made, it needs to be made in a time frame of μs to ms [44, 45]!

Mist and Fog Layers

At the next level up in the hierarchy of Fig. 2 and Table 1, that of the Mist and Fog layers, with information being shared with other vehicles and systems in the immediate vicinity, other questions arise as for instance how to allocate parking spaces to vehicles that request them, first-come, first-served or on some preferential basis.

Imagine now an autonomous vehicle in an urban area whose passenger has lost consciousness,[3] should the vehicle now attempt to get to the hospital as fast as possible, and if so what is the risk to other road users and pedestrians? In this case, actions in the Mist and Fog Layers can act to warn other users and to clear a passage for the vehicle. Here connectivity acts to reduce risk, but not without the need to address ethical issues associated, for instance, with the assessment of absolute and relative risk for the particular set of circumstances [46].

The Cloud

Cloud-based ethical concerns and issues [47, 48] are primarily with the ability to access and analyse the data collected by the system, including the risk, as discussed earlier, of algorithmic discrimination.

Trust

Studies by the World Economic Forum [49] have suggested a general lack of confidence in the way in which the internet, and by implication the Internet of Things and Cloud-based systems are both structured and operated. Here, Fig. 9 shows user responses to questions as to how their levels of trust could be increased with respect to the way in which their personal data is managed. The leading areas for change are seen to be associated with the ways in which personal data might be accessed, either by security breaches or by some form of data sharing and the ways in which such data might be used. When taken together with issues such as digital (algorithmic) discrimination, this again indicates the need for mechatronics designers, as mechatronic components and systems are in general the primary data sources, to consider privacy issues from the very beginning of the design process.

---

[3]Detected by the on-board sensors!

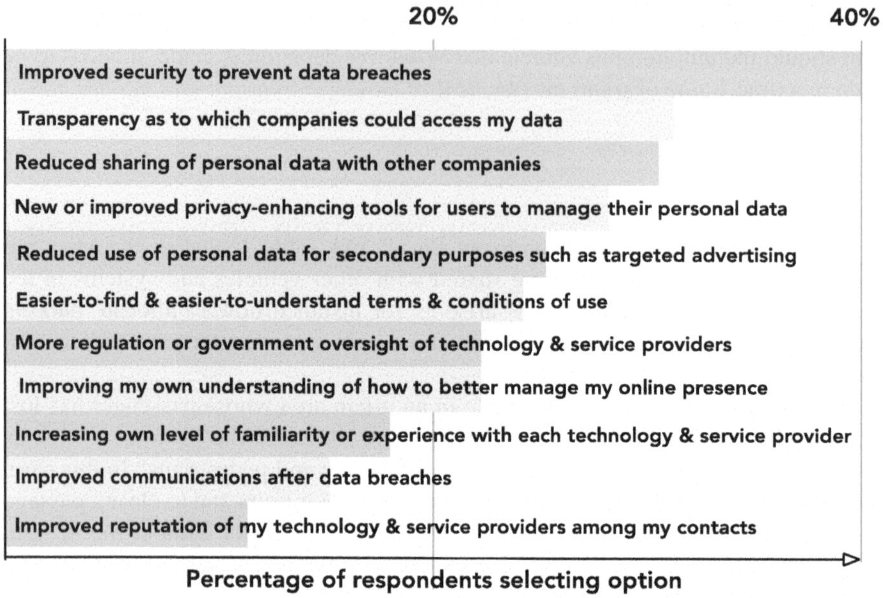

Fig. 9 User perception of changes required to improve trust of technologies and service providers {after [49]}

Conclusions

In a recent article, Xu [50] suggested that:

> Cloud computing is changing the way industries and enterprises do their businesses in that dynamically scalable and virtualized resources are provided as a service over the internet. Specifically:
>
> - Cloud computing is emerging as a major enabler for the manufacturing industry.
> - Cloud computing technologies can be adopted in manufacturing.
> - Cloud manufacturing is a pay-as-you-go business model.
> - Distributed resources are encapsulated into cloud services and managed centrally.

What is missing from this argument is that many manufacturing machines are intrinsically *mechatronic* and already include mechanical, electrical, computing and intelligent components. If machines need skilled attention to keep them running, regardless of where its data is sourced, there is little benefit to be gained from complicating what currently works! Thus while it is true that on the corporate level cloud-based systems have much to offer in regard to planning and strategy, at the operational level there may be little gain, especially if the skilled workforce has been reduced at both the factory floor and professional levels.

In looking at what is required, consider that Albert Einstein once commented that:

> We cannot solve our problems with the same thinking that we used when we created them,

so how then should mechatronics be rethought, and indeed reinvented, to accommodate the demands of Cyber-Physical Systems, the Internet of Things and The Cloud.

While a sound education in mathematics and systems engineering will always be a requirement for a mechatronics design engineer, rather than just creating "*smarter mousetraps*," perhaps research and development should be equally targeted towards usability, longevity and sustainability [51, 52]. Our throw-away society, in which defects and faults are excused and deferred to the next model or revision, unfortunately remains largely unaware of such concerns.

The paper has looked at some of the challenges facing a mechatronics in the era of Cyber-Physical Systems, the Internet of Things and Big Data and has attempted to isolate issues of concern and challenge facing systems designers, practitioners and legislators regarding privacy and ethical concerns in relation to the interaction between such systems as well as identifying contributing technical issues.

References

1. Dorogovtsev, S. N., & Mendes, J. F. (2002). Evolution of networks. *Advances in Physics, 51*(4), 1079–1187.
2. Mattern, F., & Floerkemeier, C. (2010). From the Internet of Computers to the Internet of Things. In *From active data management to event-based systems and more* (pp. 242–259). Berlin: Springer.
3. Dhamdhere, A., & Dovrolis, C. (2011). Twelve years in the evolution of the internet ecosystem. *IEEE/ACM Transactions on Networking, 19*(5), 1420–1433.
4. Foster, I., Zhao, Y., Raicu, I., & Lu, S. (2008). Cloud computing and grid computing 360-degree compared. In *IEEE Grid Computing Environments Workshop (GCE'08)* (pp. 1–10).
5. Bradley, D. A., Russell, D., Ferguson, I., Isaacs, J., & White, R. (2015). The Internet of Things—The future or the end of mechatronics. *Mechatronics, 27,* 57–74.
6. Lee, E. A., & Seshia, S. A. (2016). *Introduction to embedded systems: A cyber-physical systems approach.* MIT Press.
7. Minerva, R., Biru, A., & Rotondi, D. (2015). Towards a definition of the Internet of Things. *IEEE Internet Initiative, 1,* 1–86.
8. Rittinghouse, J. W., & Ransome, J. F. (2016). *Cloud computing: implementation, management, and security.* CRC Press.
9. Chen, H., Chiang, R. H., & Storey, V. C. (2012). Business intelligence and analytics: from big data to big impact. *MIS Quarterly*, 1165–1188.
10. Iorga, M., Feldman, L., Barton, R., Martin, M. J., Goren, N., & Mahmoudi, C. (2017). *The NIST definition of fog computing (Draft)*, NIST Special Publication 800-191.
11. Datta, S. K., Bonnet, C., & Haerri, J. (2015). Fog computing architecture to enable consumer centric Internet of Things services, *IEEE International Symposium on Consumer Electronics (ISCE)* (pp. 1–2).
12. Yi, S., Li, C., & Li, Q. (2015). A survey of fog computing: Concepts, applications and issues. In *Proceedings of the 2015 Workshop on Mobile Big Data* (pp. 37–42).
13. Gupta, H., Vahid Dastjerdi, A., Ghosh, S. K., & Buyya, R. (2017). iFogSim: A toolkit for modelling and simulation of resource management techniques in the Internet of Things. *Edge and Fog Computing Environments, Software: Practice & Experience, 47*(9), 1275–1296.
14. Munir, A., Kansakar, P., & Khan, S. U. (2017). IFCIoT: Integrated fog cloud IoT architectural paradigm for future internet of things. *IEEE Consumer Electronics Magazine, 6*(3), 74–82.

15. World Economic Forum Report Survey Report. (2015). *Deep Shift—Technology Tipping Points and Societal Impact* @ www3.weforum.org/docs/WEF_GAC15_Technological_Tipping_Points_report_2015.pdf. Accessed December 20, 2017.
16. Al-Fuqaha, A., Guizani, M., Mohammadi, M., Aledhari, M., & Ayyash, M. (2015). Internet of things: A survey on enabling technologies, protocols, and applications. *IEEE Communications Surveys & Tutorials, 17*(4), 2347–2376.
17. Botta, A., De Donato, W., Persico, V., & Pescapé, A. (2016). Integration of cloud computing and internet of things: A survey. *Future Generation Computer Systems, 56,* 684–700.
18. Weyrich, M., & Ebert, C. (2016). Reference architectures for the Internet of Things. *IEEE Software, 33*(1), 112–116.
19. Yang, Z., Yue, Y., Yang, Y., Peng, Y., Wang, X., & Liu, W. (2011). Study and application on the architecture and key technologies for IoT. *IEEE International Conference on Multimedia Technology* (pp. 747–751).
20. Wu, D., Rosen, D. W., Wang, L., & Schaefer, D. (2014). Cloud-based manufacturing: Old wine in new bottles? *Procedia CIRP, 17,* 94–99.
21. Wang, P., Gao, R. X., & Fan, Z. (2015). Cloud computing for cloud manufacturing: Benefits and limitations. *ASME Journal of Manufacturing Science and Engineering, 137*(4), 040901.
22. Facts & Data on Small Business and Entrepreneurship @ sbecouncil.org/about-us/facts-and-data/. Accessed October 4, 2018.
23. Schaar, P. (2010). Privacy by design. *Identity in the Information Society, 3*(2), 267–274.
24. Cavoukian, A., Taylor, S., & Abrams, M. E. (2010). Privacy by design: Essential for organizational accountability and strong business practices. *Identity in the Information Society, 3*(2), 405–413.
25. Radomirovic, S. (2010). Towards a model for security and privacy in the Internet of Things. In *Proceedings of the 1st International Workshop on Security of the Internet of Things*, Tokyo.
26. Weber, R. H. (2010). Internet of Things—New security and privacy challenges. *Computer Law & Security Review, 26*, 23–30.
27. British Standards Institute. (2017). PAS 185:2017 Smart Cities—Specification for establishing and implementing a security-minded approach, BSI.
28. Roman, R., Zhou, Jianying, & Lopez, J. (2013). On the features and challenges of security and privacy in distributed internet of things. *Computer Networks, 57,* 2266–2279.
29. Suo, H., Wan, J., Zou, C., & Liu, J. (2012). Security in the Internet of Things: A Review. In *Proceedings of the International Conference on Computer Science and Electronics Engineering*, (ICCSEE) (pp. 648–651).
30. Jing, Q., Vasilakos, A. V., Wan, J., Lu, J., & Qiu, D. (2014). Security of the Internet of Things: Perspectives and challenges. *Wireless Networks, 20,* 2481–2501.
31. O'Neill, C. (2017). *Weapons of math destruction*, Penguin.
32. Kambatla, K., Kollias, G., Kumar, V., & Grama, A. (2014). Trends in big data analytics. *Journal of Parallel & Distributed Computing, 74*(7), 2561–2573.
33. Chen, H., Chiang, R. H. L., & Storey, V. C. (2012). Business intelligence & analytics: From big data to big impact. *MIS Quarterly, 36*(4), 1165–1188.
34. Raghupathi, W., & Raghupathi, V. (2014). Big data analytics in healthcare: Promise and potential. *Health Information Science and Systems (Online), 2*(1), 3.
35. Lazer, D., Kennedy, R., King, G., & Vespignani, A. (2014). The parable of Google Flu: Traps in big data analysis. *Science, 343,* 1203–1205.
36. Panniello, U., Tuzhilin, A., & Gorgoglione, M. (2012). Comparing context aware recommender systems in terms of accuracy and diversity. *User Modelling & User-Adapted Interaction, 24*(1), 35–65.
37. Kroll, J. A., Barocas, S., Felten, E. W., Reidenberg, J. R., Robinson, D. G., & Yu, H. (2016). Accountable algorithms. *U P. Law. Review, 165*, 633–705.
38. Kim, P. T. (2017). Auditing algorithms for discrimination. *U. Pa Law Review Online, 166*(1), 189–203.
39. Danks, D., & London, A. J. (2017). Algorithmic bias in autonomous systems. In *Proceedings of the 26th International Joint Conference on Artificial Intelligence* (pp. 4691–4697).

40. Goodman, B. W., Flaxman, S. (2016). EU regulations on algorithmic decision-making and a "right to explanation". In *ICML Workshop Human Interpretability in Machine Learning,* New York (pp. 26–30).
41. Landau, S. (2015). Control use of Data to Protect Privacy. *Science - Special Issue The End of Privacy, 347*(6221), 504–506.
42. Watt, S., Milne, C., Bradley, D., Russell, D., Hehenberger, P., & Azorin-Lopez, J. (2016). Privacy matters-issues within mechatronics. *IFAC-PapersOnLine, 49*(21), 423–430.
43. Spangler, T. (2017). Self-driving cars programmed to decide who dies in a crash, *USA Today* @ eu.usatoday.com/story/money/cars/2017/11/23/self-driving-cars-programmed-decide-who-dies-crash/891493001/. Accessed October 4, 2018.
44. Goodall, N. J. (2016). Can you program ethics into a self-driving car? *IEEE Spectrum, 53*(6), 28–58.
45. Sparrow, R., & Howard, M. (2017). When human beings are like drunk robots: Driverless vehicles, ethics, and the future of transport. *Transportation Research Pt C: Emerging Technologies, 80,* 206–215.
46. Bonnefon, J. F., Shariff, A., & Rahwan, I. (2016). The social dilemma of autonomous vehicles. *Science, 352*(6293), 1573–1576.
47. Rogers, C., & Duranti, L. (2017). Ethics in the Cloud. *Journal of Contemporary Archival Studies, 4*(2), 1–11.
48. de Bruin, B., & Floridi, L. (2017). The ethics of cloud computing. *Science and Engineering Ethics, 23*(1), 21–39.
49. World Economic Forum White Paper. (2017). *Shaping the Future Implications of Digital Media for Society* - Valuing Personal Data and Rebuilding Trust @ www3.wefrum.org/docs/WEF_End_User_Perspective_on_Digital_Media_Survey_Summary_2017.pdf. Accessed December 20, 2017.
50. Xu, X. (2012). From cloud computing to cloud manufacturing. *Robotics & Computer Integrated Manufacturing, 28*(1), 75–86.
51. Tiefenbeck, V., Tasic, V., Schob, S., & Staake, T. (2013). Mechatronics to drive environmental sustainability: Measuring, visualizing and transforming consumer patterns on a large scale. In *IEEE Industrial Electronics Society 39th Annual Conference (IECON 2013)* (pp. 4768–4773).
52. Edwards, B. (2014). *Rough guide to sustainability: A design primer*, RIBA Publishing.

Path Planning for Semi-autonomous Agricultural Vehicles

Markus Pichler-Scheder, Reinhard Ritter, Christian Lindinger, Robert Amerstorfer and Roland Edelbauer

Abstract An on-line path planning algorithm for automated tractor steering control in greenfield farming is proposed that follows points localized on the ground, and therefore utilizes structures provided by the environment, for orientation. Points marking a swath of hay are detected using a laser rangefinder mounted on the tractor cabin. The tractor is then steered along the path so that a trailer meets the swath at its centre position. Even in the presence of outliers, the presented planning method computes the polynomial path coefficients to be used for issuing commands to the tractor. The methodology employs a multi-step initialization procedure and robust iterative optimization.

Introduction

In the past, mechatronic systems were frequently operated in a stand-alone fashion, giving the engineer full control over inputs, outputs, and many parameters and their influences. Future mechatronic designs will increasingly have to interact with their unknown and rapidly changing digital or real environment via networking, sensors, and actuators, and a much wider scope will have to be considered. A prominent example is the autonomous driving that is currently a focus of intensive research not only in the car industry but also in agriculture. In such applications, the mechatronic system must not only react to unforeseeable outside events in real-time, but also regard the environment as a symbiotic and integral element that is necessary for the mechatronic system to operate. Therefore, reinventing the paradigm of mechatronics is inevitable, offering opportunities far beyond its traditional scope.

Automation in greenfield farming is desirable for increased productivity as well as ergonomics. The process of producing and harvesting hay involves mowing and drying grass, after which the hay is arranged in rows (called '*swaths*') during the

M. Pichler-Scheder (✉) · R. Ritter
Linz Center of Mechatronics GmbH, Linz, Austria
e-mail: markus.pichler-scheder@lcm.at

C. Lindinger · R. Amerstorfer · R. Edelbauer
PÖTTINGER Landtechnik GmbH, Grieskirchen, Austria

X.-T. Yan et al. (eds.), *Reinventing Mechatronics*,
https://doi.org/10.1007/978-3-030-29131-0_4

swathing process. Afterwards, the hay is loaded onto a tractor-pulled trailer that is equipped with a rotating cylindrical pickup at its front that employs metal forks to load the hay into the trailer. For efficient and smooth loading it is essential that the pickup always meets the swath in the centre of the pickup cylinder. An automated tractor steering system is designed to issue commands to the tractor so that this demand is met, and this chapter presents the underlying path planning methodology used.

There has been extensive research on path planning for autonomous and semi-autonomous vehicles in an agricultural context. Many approaches focus on point-to-point or end-to-end planning and the avoidance of obstacles, with optimization criteria such as minimum path length, minimum energy and minimum execution time; an overview can be found in Gasparetto et al. [1] and Sorniotti et al. [2]. Path planning in autonomous driving based on polynomial, Bézier-, or B-spline representations is utilized e.g. for path smoothing and obstacle avoidance [3–5], optimization with respect to feasible driving paths [6], safe multi-vehicle operation [7], lane change on highways [8], or parallel vehicle parking [9]. An outdoor application featuring the automated optical detection of a path and the planning of vehicle navigation parameters is described in Jiang et al. [10]. While the described use case of following an unknown path in changing lighting conditions is close to the target application, the derived navigation parameters are limited to vehicle heading and offset, which does not allow for smooth and continuous steering curves for tractor control.

The algorithm proposed here operates on data points of swath positions localized by a 2-D laser rangefinder mounted on the roof of the tractor cabin that measures the terrain immediately in front of the tractor as indicated in Fig. 1. Swath positions are detected in the point cloud data, localized in a 2-dimensional global coordinate system, and entered into a point list. For localization, the current tractor position and orientation are tracked by a concurrent kinematics simulation. Due to the accumulation of tracking errors, the estimate deviates from the true tractor and trailer state over time. However, the accuracy only needs to be sufficient during the time of travel between the measurement of a particular swath point and the instant it hits the trailer pickup. The main advantages of the proposed parameterization and path planning methodology are:

- Steering curve is planned to continuously follow detected structures on a field without a pre-determined destination.
- Coordinate-system independent parameterization by the curve tangent angle as a function of path length.
- Use of a third-degree polynomial to facilitate S-shaped curves.
- Direct conversion of the polynomial tangent angle to the required steering control curves.
- Incorporation of feasibility constraints given by the current tractor-trailer and steering curvature configuration.
- On-line planning of continuous steering curves as new curve points are added and old points removed.
- Robust least-squares method insensitive to outlier measurements.

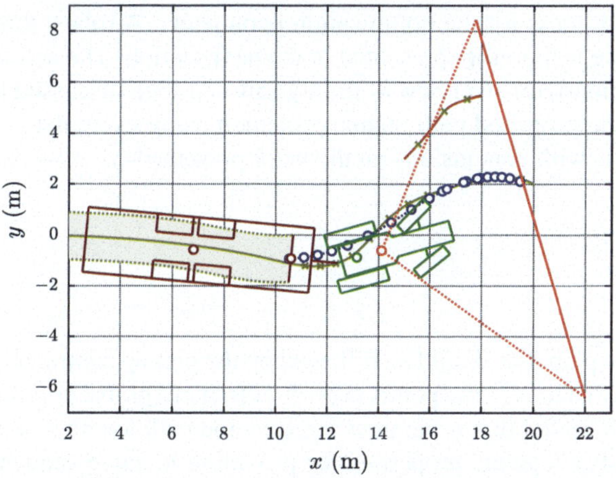

Fig. 1 Top view of tractor and trailer combination with the current pickup centre position marked with a red circle. Swath is shown as a solid brown line, while the area covered by the pickup is shown as a light brown area with dotted margins. Rays from the laser scanner mounted on the tractor cabin intersecting the ground plane are shown as a light red line. Measured data points marked as brown × including outlier measurements on an undesired curve shown in dark red. Data points serve as the basis for planning the travelled curve, which is shown as a blue dotted line with blue circles starting from the pickup position at the time of planning and extending up to the furthest measured point

This is at the cost of sufficient in-vehicle computing power which can, however, be easily provided, for instance, by an embedded Linux board.

The chapter builds on and develops previous work [11] by the authors which contained the description of the presented path planning algorithm, with additional simulation results for algorithm behaviour in challenging situations.[1] Instead of relying on data from a satellite navigation system, the methodology is able to assess the situation at the time of loading and does not require to previously map datapoints during the swathing process. In an effort to utilize structures provided by the environment as a necessary system component for localization and orientation estimation, the scope of this work is wider than the traditional field of mechatronics.

The chapter is arranged as follows: A polynomial path model used for planning and steering control is presented and a power series representation of the path position depending on the travelled path length is derived to facilitate computation. Since the planned path curve should be as close as possible to the data points, an optimization algorithm for computing path coefficients is required. The method must be robust to the erroneous outlier measurements that are inevitable in the given outdoor scenario. The following section therefore briefly summarizes a supporting method for robust optimization. The typically employed iterative methods for non-linear optimization such as the Newton-, Gauss-Newton, or Levenberg-Marquardt-algorithms [12] all

[1] The chapter also contains corrections for some minor errata in [11].

require an initial guess for the optimization parameters. A robust three-step initialization procedure is therefore presented in the next section. The actual optimization including a summarizing overview of the algorithm is then discussed followed by an application of the proposed path planning method using exemplary measurements. The chapter ends with conclusions on the work presented.

Path Model

For planning the path that should be followed by the pickup centre, it is assumed that operation is on a list of N_V data points $(x_V[n], y_V[n])$ in a global coordinate system. A polynomial path model that is parameterized by the path length s is then used. This choice offers advantageous properties for providing a stable sequence of steering commands to the tractor, an elaboration on this issue is, however, out of scope of this chapter. Further, a third order polynomial allows the steering of S-shaped curves which are necessary when the initial tractor position is off to the side of the sequence of data points. The starting path length at the time of planning is denoted as s_0, which is the total path length travelled since system startup. As an unambiguous expression for the path depending on $\tilde{s} \equiv s - s_0$ the path-dependent tangent angle to the curve ψ in every point is used,

$$\psi(\mathbf{a}, \tilde{s}) = a_0 + a_1\tilde{s} + a_2\tilde{s}^2 + a_3\tilde{s}^3 \quad \text{with} \quad \mathbf{a} = \begin{bmatrix} a_0 \; a_1 \; a_2 \; a_3 \end{bmatrix}^{\mathrm{T}} \tag{1}$$

where the (partly) unknown coefficients that will be the result of the path planning are assembled in the vector $\mathbf{a}$.

A differential change in position $\mathrm{d}x_C$ and $\mathrm{d}y_C$ in x- and y-directions is given depending on the differential path length $\mathrm{d}s$ as

$$\mathrm{d}x_C = \cos(\psi)\mathrm{d}s \quad \mathrm{d}y_C = \sin(\psi)\mathrm{d}s \tag{2}$$

For the position on the path $(x_C(\mathbf{a}, \tilde{s}), y_C(\mathbf{a}, \tilde{s}))$ after a traveled path length $\tilde{s}$ it is necessary to integrate (2) in $\mathrm{d}s$. Equations (3) and (4) are then derived from the starting point of the pickup $(x_C(s_0), y_C(s_0))$ as

$$x_C(\mathbf{a}, \tilde{s}) = x_C(s_0) + \int_0^{\tilde{s}} \cos(\psi(\mathbf{a}, \sigma))\mathrm{d}\sigma \tag{3}$$

$$y_C(\mathbf{a}, \tilde{s}) = y_C(s_0) + \int_0^{\tilde{s}} \sin(\psi(\mathbf{a}, \sigma))\mathrm{d}\sigma. \tag{4}$$

Both the initial position $(x_C(s_0), y_C(s_0))$ and the initial angle a_0 are known from the tracked current tractor and trailer positions and orientations. A coordinate transformation of the initial pickup position to the origin is therefore performed, with the initial direction of movement along the positive x-axis. Hence

$$\begin{bmatrix} \tilde{x}_C(\mathbf{a}, \tilde{s}) \\ \tilde{y}_C(\mathbf{a}, \tilde{s}) \end{bmatrix} \equiv \begin{bmatrix} \cos(a_0) & \sin(a_0) \\ -\sin(a_0) & \cos(a_0) \end{bmatrix} \begin{bmatrix} x_C(\mathbf{a}, \tilde{s}) - x_C(s_0) \\ y_C(\mathbf{a}, \tilde{s}) - y_C(s_0) \end{bmatrix} \tag{5}$$

$$\begin{bmatrix} \tilde{x}_V[n] \\ \tilde{y}_V[n] \end{bmatrix} \equiv \begin{bmatrix} \cos(a_0) & \sin(a_0) \\ -\sin(a_0) & \cos(a_0) \end{bmatrix} \begin{bmatrix} x_V[n] - x_C(s_0) \\ y_V[n] - y_C(s_0) \end{bmatrix} \tag{6}$$

Inserting (1) into (3) and (4) and then in (5) yields

$$\tilde{x}_C(\mathbf{a}, \tilde{s}) = \int_0^{\tilde{s}} \cos\left(a_1\sigma + a_2\sigma^2 + a_3\sigma^3\right) \mathrm{d}\sigma \tag{7}$$

$$\tilde{y}_C(\mathbf{a}, \tilde{s}) = \int_0^{\tilde{s}} \sin\left(a_1\sigma + a_2\sigma^2 + a_3\sigma^3\right) \mathrm{d}\sigma \tag{8}$$

which cannot be solved in closed form.

It is therefore proposed to use a power series representation for the cos- and sin-functions,

$$\cos(x) = \lim_{N\to\infty} \sum_{n=0}^{N} \frac{(-1)^n}{(2n)!} x^{2n} \quad \sin(x) = \lim_{N\to\infty} \sum_{n=0}^{N} \frac{(-1)^n}{(2n+1)!} x^{2n+1} \tag{9}$$

that provide a good approximation even if the infinite sum is truncated to a relatively low number. Since in a practical scenario the path angle in the transformed coordinate system will never exceed $\pi/2$, the approximation for $N = 3$ will be sufficiently accurate.

Using the multinomial decomposition

$$(x_1 + x_2 + x_3)^n = \sum_{k_1+k_2+k_3=n} \frac{n!}{k_1!k_2!k_2!} \prod_{m=1}^{3} x_m^{k_m} \tag{10}$$

it is possible to write the integrands in (7) and (8) as polynomials in the variable σ,

$$\begin{aligned} \cos\left(a_1\sigma + a_2\sigma^2 + a_3\sigma^3\right) = \sum_{n=0}^{\infty} \frac{(-1)^n}{(2n)!} \sum_{k_2=0}^{2n} \sum_{k_3=0}^{2n-k_2} \frac{(2n)!}{k_2!k_3!(2n-k_2-k_3)!} \\ \cdot a_1^{2n-k_2-k_3} a_2^{k_2} a_3^{k_3} \sigma^{2n+k_2+2k_3} \end{aligned} \tag{11}$$

$$\begin{aligned} \sin\left(a_1\sigma + a_2\sigma^2 + a_3\sigma^3\right) = \sum_{n=0}^{\infty} \frac{(-1)^n}{(2n+1)!} \sum_{k_2=0}^{2n+1} \sum_{k_3=0}^{2n+1-k_2} \frac{(2n+1)!}{k_2!k_3!(2n+1-k_2-k_3)!} \\ \cdot a_1^{2n-k_2-k_3} a_2^{k_2} a_3^{k_3} \sigma^{2n+k_2+2k_3}. \end{aligned} \tag{12}$$

These expressions can be solved using the monomial integration rule

$$\int_0^{\tilde{s}} \sigma^k \mathrm{d}\sigma = \frac{\tilde{s}^{k+1}}{k+1}, \tag{13}$$

to obtain a power series representation for the pickup position $(\tilde{x}_\mathrm{C}(\mathbf{a}, \tilde{s}), \tilde{y}_\mathrm{C}(\mathbf{a}, \tilde{s}))$ depending on polynomial coefficients $\mathbf{a}$ as a function of path length $\tilde{s}$ as given in Eqs. (14) and (15). In the following, the coefficients $\mathbf{a}$ will be optimized so that this traveled curve is as close as possible to the measured data points $(\tilde{x}_\mathrm{V}[n], \tilde{y}_\mathrm{V}[n])$.

$$\begin{aligned}\tilde{x}_\mathrm{C}(\mathbf{a}, \tilde{s}) = \sum_{n=0}^{\infty} \frac{(-1)^n}{(2n)!} \sum_{k_2=0}^{2n} \sum_{k_3=0}^{2n-k_2} & \frac{(2n)!}{k_2! k_3! (2n-k_2-k_3)!} \\ & \cdot \frac{a_1^{2n-k_2-k_3} a_2^{k_2} a_3^{k_3} \tilde{s}^{2n+1+k_2+2k_3}}{2n+1+k_2+2k_3}\end{aligned} \tag{14}$$

$$\begin{aligned}\tilde{y}_\mathrm{C}(\mathbf{a}, \tilde{s}) = \sum_{n=0}^{\infty} \frac{(-1)^n}{(2n+1)!} \sum_{k_2=0}^{2n+1} \sum_{k_3=0}^{2n+1-k_2} & \frac{(2n+1)!}{k_2! k_3! (2n+1-k_2-k_3)!} \\ & \cdot \frac{a_1^{2n+1-k_2-k_3} a_2^{k_2} a_3^{k_3} \tilde{s}^{2n+2+k_2+2k_3}}{2n+2+k_2+2k_3}\end{aligned} \tag{15}$$

Robust Estimation

To obtain robust estimates for path curves even in the presence of outliers, an iteratively reweighted least squares method is employed [13]. For a parameterized model to be computed from data containing erroneous measurements, this method assigns weights $w[n]$ to individual data points based on their residual with respect to the model. Initially, all weights are set to $w[n] = 1$ and an initial set of model parameters must be given. If $r[n]$ denotes the current residual of a data point from the estimated model, then a robust estimate $\hat{\sigma}$ for the standard deviation of the residual can be obtained as the median absolute deviation (MAD) which is computed as:

$$\hat{\sigma} = \frac{1}{0.6745} \operatorname*{median}_n |r[n]|. \tag{16}$$

The factor of $1/0.6745$ leads to an estimate equal to the standard deviation if the residual is normally distributed. Weights $w[n]$ for the individual data points n can be computed using Tukey's bisquare-function [14]:

$$w[n] = \begin{cases} \left(1 - \left(\frac{r[n]}{c \cdot \hat{\sigma}}\right)^2\right)^2 & \text{if} |r[n]| < c \cdot \hat{\sigma} \text{ with } c = 4.685 \\ 0 & \text{else.} \end{cases} \tag{17}$$

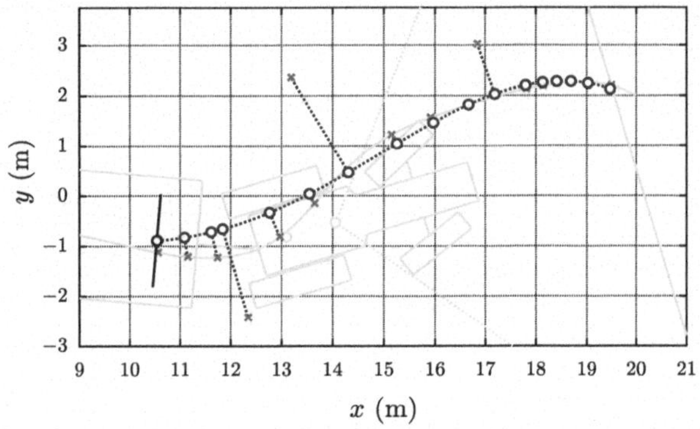

Fig. 2 Example for robust curve fitting in the presence of outliers which should only be minimally affected by erroneous measurements off the swath curve

Thus, a data point gets a larger weight if its residual with respect to the model is smaller. If the residual exceeds the value of $c \cdot \hat{\sigma}$, then the data point is treated as an outlier and its weight becomes zero. In an iterative procedure, the weights are used for computing an improved estimate of the model parameters. Data points with weight zero are not considered in this iteration. The method is repeated until it converges. Using this method, curve fitting is largely unaffected by outlier measurements as indicated in the example shown in Fig. 2.

Initialization

Any iterative optimization procedure must be initialized with model parameters that are near to the optimum so that the optimization converges. In particular in the presence of outliers, this can be a challenging task, as there is no generally valid assumption about the properties of the optimization function (such as convexity). A three-step initialization for the path planning procedure is therefore proposed. In conjunction with the detection of divergence of the iterative optimization that may happen in rare cases, this method is robust in practical situations, as will be shown by simulation results in the appropriate section of this chapter.

Straight Line Estimate

In the first step a straight line through the origin is fitted to the data points. Although this model will usually be a very inaccurate representation of the final curve, it can still be used to identify large outliers and exclude them in the next step by determining their perpendicular distance to the line. As an estimate of the angle $\hat{\alpha}$ of the line the median of the individual angles with respect to the $\tilde{x}$-axis can be used,

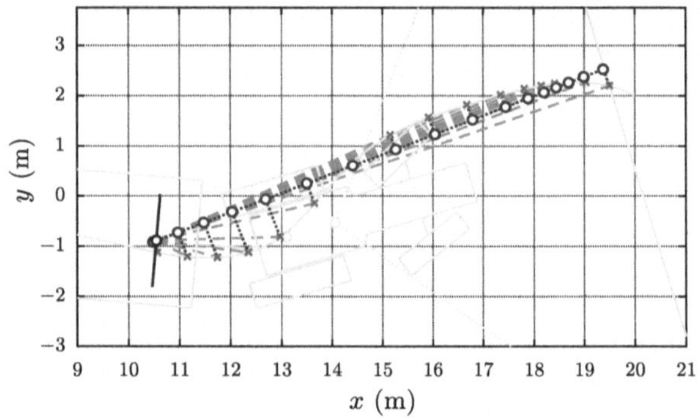

Fig. 3 Example of a straight line estimate. Lines from the pickup position to the individual data points are shown as grey dashed lines, the straight line estimate with median angle is shown as a blue line with blue circles indicating the points on the line corresponding to data points

$$\hat{\alpha} = \underset{n}{\text{median}}(\arctan(\tilde{y}_\text{V}[n], \tilde{x}_\text{V}[n])) \tag{18}$$

where arctan denotes the two-argument arc tangent function. With the breakdown point of the median at 50%, at most $\frac{N_\text{V}}{2}$ outliers with arbitrary magnitude will not affect the result. The residuals of the individual points are obtained as their projection onto the perpendicular unit vector, i.e.

$$r_\text{s}[n] = \begin{bmatrix} -\sin(\hat{\alpha}) \\ \cos(\hat{\alpha}) \end{bmatrix}^\text{T} \begin{bmatrix} \tilde{x}_\text{V}[n] \\ \tilde{y}_\text{V}[n] \end{bmatrix} \tag{19}$$

Using Eqs. (16) and (17) it is possible to obtain a first set of weights $w_\text{s}[n]$ for the next step. This procedure is illustrated in Fig. 3. Lines from the pickup position to the individual data points are shown as grey dashed lines, the line with median angle $\hat{\alpha}$ that is used for the straight line estimate is shown in blue with blue circles.

Circle Estimate

In the second initialization step, the path curve is approximated by a circle through the origin that has its centre on the $\tilde{y}$-axis. As fitting a circle would be a non-linear optimization problem in itself, an approximation by a parabola of the form of Eq. (20) is used:

$$\tilde{y}_\text{C} = \frac{\tilde{a}_1}{2}\tilde{x}_\text{C}^2. \tag{20}$$

Using the weights from the first initialization step, the parabola is fitted using the weighted least squares criterion by setting the derivative with respect to the model parameter $\tilde{a}_1$ to zero, i.e.,

$$\begin{aligned}\hat{\tilde{a}}_1 &= \arg\min_{\tilde{a}_1} \sum_{n=0}^{N_V-1} w_s[n]\left(\tilde{y}_V[n] - \frac{\tilde{a}_1}{2}\tilde{x}_V^2[n]\right)^2 \\ &= \frac{2\sum_{n=0}^{N_V-1} w_s[n] y_V[n] x_V^2[n]}{\sum_{n=0}^{N_V-1} w_s[n] x_V^4[n]}.\end{aligned} \tag{21}$$

The approximation of a circle by the curvature of a parabola is valid if the data points are distributed along a circular arc of not more than about 1/8 of a full circle, i.e., if $\arctan(|\tilde{y}_V[n]|, \tilde{x}_V[n]) < \frac{\pi}{8}$. Figure 4 shows the result of a circular arc (blue line) with curvature identical to an estimated parabola (grey dashed line). For initialization, the induced systematic error is below 0.5 m at the end of the curve and thus sufficiently small. The associated path lengths $\tilde{s}_c[n]$ to the points on the circle closest to the data points are computed as

$$\tilde{s}_c[n] = \begin{cases} \tilde{x}_V[n] & \text{for } \tilde{a}_1 = 0 \\ \frac{1}{|\tilde{a}_1|}\arctan\left(\tilde{x}_V[n], \operatorname{sgn}(\tilde{a}_1)\left(\frac{1}{\tilde{a}_1} - \tilde{y}_V[n]\right)\right) & \text{else} \end{cases} \tag{22}$$

with corresponding circle points $(\tilde{x}_{C1}[n], \tilde{y}_{C1}[n])$

$$\begin{aligned}\tilde{x}_{C1}[n] &= \begin{cases} \tilde{s}_c[n] & \text{for } \tilde{a}_1 = 0 \\ \frac{\sin(\tilde{a}_1\tilde{s}_c[n])}{\tilde{a}_1} & \text{else} \end{cases} \\ \tilde{y}_{C1}[n] &= \begin{cases} \tilde{s}_c[n] & \text{for } \tilde{a}_1 = 0 \\ \frac{1-\cos(\tilde{a}_1\tilde{s}_c[n])}{\tilde{a}_1} & \text{else} \end{cases}\end{aligned} \tag{23}$$

The residuals are computed as the normal projection of the data points onto the circle, i.e.,

$$r_c[n] = \begin{bmatrix} -\sin(\tilde{a}_1\tilde{s}_c[n]) \\ \cos(\tilde{a}_1\tilde{s}_c[n]) \end{bmatrix}^T \begin{bmatrix} \tilde{x}_V[n] - \tilde{x}_{C1}[n] \\ \tilde{y}_V[n] - \tilde{y}_{C1}[n] \end{bmatrix} \tag{24}$$

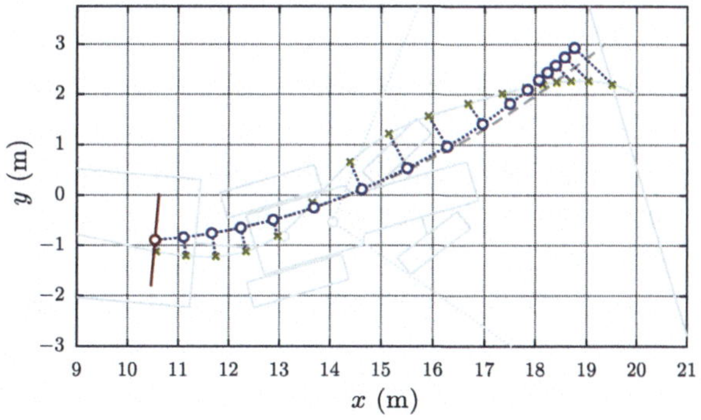

Fig. 4 Example of a circle estimate. The fitted parabola is shown as a grey dashed line, the circle estimate with identical curvature is shown in blue with blue circles indicating the points on the circle corresponding to the data points

A new set of weights $w_c[n]$ for the next step is again computed using Eqs. (16) and (17).

Linear Polynomial Coefficient Estimate

In the third step, the path curve is linearized about the circle with curvature $\tilde{a}_1$ estimated in the last step. Parameter a_1 is then split as $\tilde{a}_1$ and a remainder Δa_1 such that

$$a_1 = \tilde{a}_1 + \Delta a_1 \tag{25}$$

which allows the argument of the sin- and cos-functions in Eqs. (7) and (8) to be split as

$$a_1\sigma + a_2\sigma^2 + a_3\sigma^3 = \tilde{a}_1\sigma + \Delta a_1\sigma + a_2\sigma^2 + a_3\sigma^3. \tag{26}$$

For a linear approximation it is assumed that the estimated circular arc $\tilde{a}_1$ is dominant, which is reasonable considering the typical shape of hey rows as they occur in the field (cf. [11]). It can therefore be assumed that the deviation from this arc is small, i.e.,

$$\Delta a_1\sigma + a_2\sigma^2 + a_3\sigma^3 \ll 1 \text{ for } \sigma \in \left[0, \tilde{s}[n]\right] \quad \forall n, \tag{27}$$

then, using

$$\cos(x) \approxeq 1 \ \sin(x) \approxeq x \ \text{ for } x \ll 1 \tag{28}$$

Equations (7) and (8) can be approximated as

$$\tilde{x}_C(\mathbf{a}, \tilde{s}) \approxeq \int_0^{\tilde{s}} \cos(\tilde{a}_1\sigma)d\sigma - \int_0^{\tilde{s}} \sin(\tilde{a}_1\sigma)\left(\Delta a_1\sigma + a_2\sigma^2 + a_3\sigma^3\right)d\sigma \tag{29}$$

$$\tilde{y}_C(\mathbf{a}, \tilde{s}) \approxeq \int_0^{\tilde{s}} \sin(\tilde{a}_1\sigma)d\sigma + \int_0^{\tilde{s}} \cos(\tilde{a}_1\sigma)\left(\Delta a_1\sigma + a_2\sigma^2 + a_3\sigma^3\right)d\sigma, \tag{30}$$

which can easily be solved analytically and, after inserting $\tilde{s}_c[n]$, takes the form of a linear function in the elements of $\mathbf{a}$,

$$\tilde{x}_C(\mathbf{a}, \tilde{s}_c[n]) \approxeq \tilde{x}_{C1}[n] + \tilde{x}_{\Delta C1}[n]\Delta a_1 + \tilde{x}_{C2}[n]a_2 + \tilde{x}_{C3}[n]a_3 \tag{31}$$

$$\tilde{y}_C(\mathbf{a}, \tilde{s}_c[n]) \approxeq \tilde{y}_{C1}[n] + \tilde{y}_{\Delta C1}[n]\Delta a_1 + \tilde{y}_{C2}[n]a_2 + \tilde{y}_{C3}[n]a_3 \tag{32}$$

The (lengthy) results for the coefficients $\tilde{x}/\tilde{y}_{\Delta C1/C2/C3}$ are computed in a straightforward manner and will not be listed here. The residual between the circular arc and the polynomial curve is obtained by projection on the arc normal vector, i.e.

$$\begin{aligned} r_{\mathrm{p}}[n] &\equiv \begin{bmatrix} -\sin(\tilde{a}_1\tilde{s}_{\mathrm{c}}[n]) \\ \cos(\tilde{a}_1\tilde{s}_{\mathrm{c}}[n]) \end{bmatrix}^{\mathrm{T}} \cdot \begin{bmatrix} \tilde{x}_{\mathrm{C}}(\mathbf{a}, \tilde{s}_{\mathrm{c}}[n]) - \tilde{x}_{\mathrm{C1}}[n] \\ \tilde{y}_{\mathrm{C}}(\mathbf{a}, \tilde{s}_{\mathrm{c}}[n]) - \tilde{y}_{\mathrm{C1}}[n] \end{bmatrix} \\ &= b_1[n]\Delta a_1 + b_2[n]a_2 + b_3[n]a_3 = r_{\mathrm{c}}[n] \end{aligned} \tag{33}$$

where

$$b_1[n] \equiv \begin{cases} \frac{\tilde{s}_{\mathrm{c}}^2[n]}{2} & \text{for } \tilde{a}_1 = 0 \\ \frac{1-\cos(\tilde{a}_1\tilde{s}_{\mathrm{c}}[n])}{\tilde{a}_1^2} & \text{else} \end{cases} \tag{34}$$

$$b_2[n] \equiv \begin{cases} \frac{\tilde{s}_{\mathrm{c}}^3[n]}{3} & \text{for } \tilde{a}_1 = 0 \\ \frac{2(\tilde{a}_1\tilde{s}_{\mathrm{c}}[n]-\sin(\tilde{a}_1\tilde{s}_{\mathrm{c}}[n]))}{\tilde{a}_1^3} & \text{else} \end{cases} \tag{35}$$

$$b_3[n] \equiv \begin{cases} \frac{\tilde{s}_{\mathrm{c}}^4[n]}{4} & \text{for } \tilde{a}_1 = 0 \\ \frac{-6(1-\cos(\tilde{a}_1\tilde{s}_{\mathrm{c}}[n]))+3\tilde{a}_1^2\tilde{s}_{\mathrm{c}}^2[n]}{\tilde{a}_1^4} & \text{else} \end{cases} \tag{36}$$

For $n \in [0, N_{\mathrm{V}} - 1]$, Eq. (33) yields a linear system of equations, which can be solved for the desired coefficients Δa_1, a_2, and a_3. In reality, coefficient a_1 cannot be chosen independently for stability reasons, the discussion of which is out of scope of this work. Therefore, observing the dependency

$$a_1 = \tilde{a}_1 + \Delta a_1 = a_{10} + a_{12}a_2 + a_{13}a_3 \tag{37}$$

with known coefficients a_{10}, a_{12}, and a_{13}, and including the weights computed in the last step, the linear system can be written as:

$$\mathbf{W}_{\mathrm{c}}\tilde{\mathbf{B}}\tilde{a}_{\mathrm{p}} = \mathbf{W}_{\mathrm{c}}\tilde{\mathbf{r}} \tag{38}$$

with

$$\tilde{\mathbf{B}} \equiv \begin{bmatrix} \tilde{b}_2[0] & \tilde{b}_3[0] \\ \vdots & \vdots \\ \tilde{b}_2[N_{\mathrm{V}}-1] & \tilde{b}_3[N_{\mathrm{V}}-1] \end{bmatrix} \quad \tilde{\mathbf{r}} \equiv \begin{bmatrix} \tilde{r}[0] \\ \vdots \\ \tilde{r}[N_{\mathrm{V}}-1] \end{bmatrix} \tag{39}$$

$$\mathbf{W}_{\mathrm{c}} \equiv \mathrm{diag}(\mathbf{w}_{\mathrm{c}}) \ \mathbf{w}_{\mathrm{c}} = \begin{bmatrix} w_c[0] \\ \vdots \\ w_c[N_{\mathrm{V}}-1] \end{bmatrix} \quad \tilde{\mathbf{a}}_{\mathrm{p}} = \begin{bmatrix} a_{\mathrm{p2}} \\ a_{\mathrm{p3}} \end{bmatrix} \tag{40}$$

where

$$\tilde{b}_2[n] = b_2[n] + b_1[n]a_{12} \tag{41}$$

$$\tilde{b}_3[n] = b_3[n] + b_1[n]a_{13} \tag{42}$$

$$\tilde{r}[n] = \tilde{r}_c[n] + b_1[n](\tilde{a}_1 - a_{10}). \tag{43}$$

The linear system of equations (38) is solved for $\tilde{\mathbf{a}}_p$

$$\hat{\tilde{\mathbf{a}}}_p = \left(\mathbf{W}_c\tilde{\mathbf{B}}\right)^+ \mathbf{W}_c\tilde{\mathbf{r}}, \tag{44}$$

where $\cdot^+$ denotes the Moore-Penrose pseudo-inverse, to provide an initialization for the iterative optimizer.

Figure 5 shows an example of the result of linear estimation of the polynomial coefficients. Even though the curve is S-shaped, the linear estimate already corresponds well to the final estimated polynomial (cf. Figure 6). However, due to the approximate nature and the use of path lengths from the foregoing circle estimate, the points on the curve (blue circles) corresponding to the data points are not the closest points on the curve.

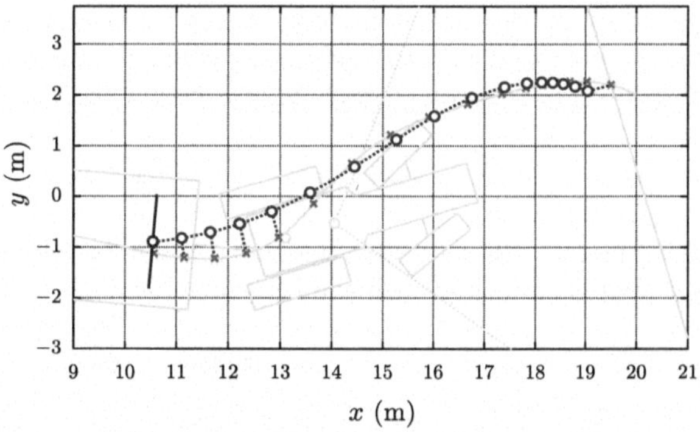

Fig. 5 Example of a linear polynomial coefficient estimate shown as a blue line with blue circles indicating the points on the polynomial curve corresponding to the data points. Due to the linear approximation, the blue circles are not the curve points closest to the data points

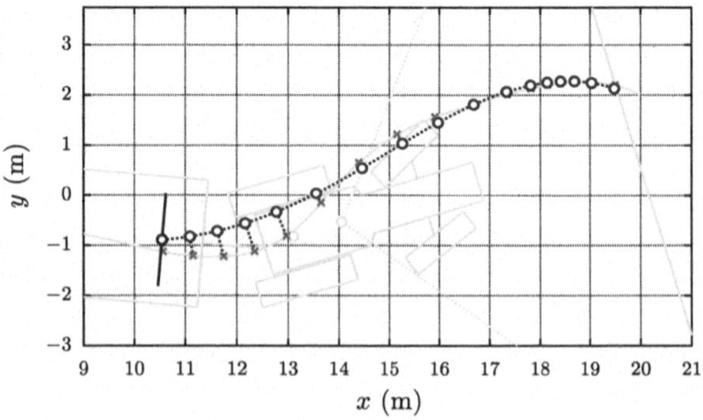

Fig. 6 Example of a polynomial curve estimate that is iteratively fitted to the data points in a least-squares sense

Path Planning

The elements of the coefficient vector $\tilde{\mathbf{a}} = \begin{bmatrix} a_2 \; a_3 \end{bmatrix}^{\mathrm{T}}$ will now be computed so that the planned path curve fits the data points $(\tilde{x}_{\mathrm{V}}[n], \tilde{y}_{\mathrm{V}}[n])$ best in a weighted least squares sense, i.e.,

$$\hat{\mathbf{a}} = \arg\min_{\mathbf{a}} f_{\mathbf{a}}(\mathbf{a}) \text{ with}$$
$$f_{\mathbf{a}}(\mathbf{a}) \equiv \sum_{n=0}^{N_{\mathrm{V}}-1} w[n]\big(\Delta\tilde{x}^2(\mathbf{a}, \tilde{s}[n]) + \Delta\tilde{y}^2(\mathbf{a}, \tilde{s}[n])\big), \tag{45}$$

where

$$\Delta\tilde{x}(\mathbf{a}, \tilde{s}[n]) \equiv \tilde{x}_{\mathrm{V}}[n] - \tilde{x}_{\mathrm{C}}(\mathbf{a}, \tilde{s}[n]) \; \Delta\tilde{y}(\mathbf{a}, \tilde{s}[n]) \equiv \tilde{y}_{\mathrm{V}}[n] - \tilde{y}_{\mathrm{C}}(\mathbf{a}, \tilde{s}[n]). \tag{46}$$

This nonlinear optimization requires the gradient of $f_{\mathbf{a}}(\mathbf{a})$ (45) w.r.t. $\tilde{\mathbf{a}}$ to be zero,

$$\begin{aligned} \mathbf{g}_{\tilde{\mathbf{a}}}(a) &\equiv \frac{\partial f_{\mathbf{a}}}{\partial \tilde{\mathbf{a}}}(\mathbf{a}) \\ &= 2\sum_{n=0}^{N_{\mathrm{V}}-1} w[n]\left(\Delta\tilde{x}(\mathbf{a}, \tilde{s}[n])\frac{\partial \tilde{x}_{\mathrm{C}}}{\partial \tilde{\mathbf{a}}}(\mathbf{a}, \tilde{s}[n]) + \Delta\tilde{y}(\mathbf{a}, \tilde{s}[n])\frac{\partial \tilde{y}_{\mathrm{C}}}{\partial \tilde{\mathbf{a}}}(\mathbf{a}, \tilde{s}[n])\right) = 0, \end{aligned} \tag{47}$$

which can be solved e.g. using Newton's algorithm that requires the Hessian

$$\begin{aligned} \mathbf{H}_{\tilde{\mathbf{a}}}(\mathbf{a}) \equiv \frac{\partial \mathbf{g}_{\tilde{\mathbf{a}}}}{\partial \tilde{\mathbf{a}}^T}(\mathbf{a}) = 2\sum_{n=0}^{N_{\mathrm{V}}-1} w[n] \\ \left(\left(\frac{\partial \tilde{x}_{\mathrm{C}}}{\partial \tilde{\mathbf{a}}}(\mathbf{a}, \tilde{s}[n])\right)\left(\frac{\partial \tilde{x}_{\mathrm{C}}}{\partial \tilde{\mathbf{a}}}(\mathbf{a}, \tilde{s}[n])\right)^{\mathrm{T}} + \Delta\tilde{x}(\mathbf{a}, \tilde{s}[n])\frac{\partial \tilde{x}_{\mathrm{C}}}{\partial \tilde{\mathbf{a}}\partial \tilde{\mathbf{a}}^T}(\mathbf{a}, \tilde{s}[n])\right. \\ \left.\left(\frac{\partial \tilde{y}_{\mathrm{C}}}{\partial \tilde{\mathbf{a}}}(\mathbf{a}, \tilde{s}[n])\right)\left(\frac{\partial \tilde{y}_{\mathrm{C}}}{\partial \tilde{\mathbf{a}}}(\mathbf{a}, \tilde{s}[n])\right)^{\mathrm{T}} + \Delta\tilde{y}(\mathbf{a}, \tilde{s}[n])\frac{\partial \tilde{y}_{\mathrm{C}}}{\partial \tilde{\mathbf{a}}\partial \tilde{\mathbf{a}}^T}(\mathbf{a}, \tilde{s}[n])\right) \end{aligned} \tag{48}$$

The partial derivatives of $\tilde{x}_{\mathrm{C}}$ and $\tilde{y}_{\mathrm{C}}$ according to Eqs. (7) and (8) in Eqs. (47) and (48) contain products of powers of the coefficients a_k which must be computed using the product rule and the monomial rule of differentiation

$$\frac{\partial}{\partial a_j} a_i^m = \begin{cases} 0 & \text{for } m = 0 \\ m a_i^{m-1} \frac{\partial}{\partial a_j} a_i & \text{else} \end{cases} \tag{49}$$

$$\frac{\partial^2}{\partial a_j \partial a_k} a_i^m = \begin{cases} 0 & \text{for } m \leq 1 \\ m(m-1) a_i^{m-2}\left(\frac{\partial}{\partial a_j} a_i\right)\left(\frac{\partial}{\partial a_k} a_i\right) + m a_i^{m-1} \frac{\partial^2}{\partial a_j \partial a_k} a_i & \text{else} \end{cases} \tag{50}$$

Note that the dependency of Eq. (37) must be observed.

As a condition on the path length $\tilde{s}[n]$ up to the curve point $(\tilde{x}_C(\mathbf{a}, \tilde{s}[n]), \tilde{y}_C(\mathbf{a}, \tilde{s}[n]))$ corresponding to data point $(\tilde{x}_V[n], \tilde{y}_V[n])$ it is required that those are the points on the curve closest to the data points. This translates to the condition that the vector between the curve after length $\tilde{s}[n]$ and the associated data point is perpendicular to the curve tangent. The tangent vector is given by the derivative of the curve coordinates with respect to the path length, hence

$$\begin{bmatrix} \frac{\partial \tilde{x}_C}{\partial \tilde{s}}(\mathbf{a}, \tilde{s}[n]) \\ \frac{\partial \tilde{y}_C}{\partial \tilde{s}}(\mathbf{a}, \tilde{s}[n]) \end{bmatrix}^T \cdot \begin{bmatrix} \Delta\tilde{x}(\mathbf{a}, \tilde{s}[n]) \\ \Delta\tilde{y}(\mathbf{a}, \tilde{s}[n]) \end{bmatrix} = 0 \tag{51}$$

or

$$g_{\tilde{s},n}(\tilde{s}[n]) \equiv \Delta\tilde{x}(\mathbf{a}, \tilde{s}[n]) \frac{\partial \tilde{x}_C}{\partial s}(\mathbf{a}, \tilde{s}[n]) + \Delta\tilde{y}(\mathbf{a}, \tilde{s}[n]) \frac{\partial \tilde{y}_C}{\partial s}(\mathbf{a}, \tilde{s}[n]) = 0. \tag{52}$$

The derivative of $g_{\tilde{s},n}$ is given by

$$\begin{aligned} H_{\tilde{s},n}(\tilde{s}[n]) \equiv \frac{\partial g_{\tilde{s},n}}{\partial \tilde{s}}(\tilde{s}[n]) &= \left(\frac{\partial \tilde{x}_C}{\partial \tilde{s}}(\mathbf{a}, \tilde{s}[n])\right)^2 \\ &+ \Delta\tilde{x}(\mathbf{a}, \tilde{s}[n]) \frac{\partial^2 \tilde{x}_C}{\partial \tilde{s}^2}(\mathbf{a}, \tilde{s}[n]) \\ &+ \left(\frac{\partial \tilde{y}_C}{\partial \tilde{s}}(\mathbf{a}, \tilde{s}[n])\right)^2 + \Delta\tilde{y}(\mathbf{a}, \tilde{s}[n]) \frac{\partial^2 \tilde{y}_C}{\partial \tilde{s}^2}(\mathbf{a}, \tilde{s}[n]) \end{aligned} \tag{53}$$

Starting from the initialization of $\tilde{\mathbf{a}}^{[0]} \equiv \hat{\tilde{\mathbf{a}}}_p$, $\mathbf{w}^{[0]} \equiv \mathbf{w}_c$, and $\tilde{s}[n]^{[0]} \equiv \tilde{s}_c[n]$ a two-step optimization is proposed for efficiency by first computing an update to the polynomial parameters $\tilde{\mathbf{a}}$

$$\delta\tilde{\mathbf{a}}^{[k]} = -\mathbf{H}_{\tilde{\mathbf{a}}}^{-1}\left(\tilde{\mathbf{a}}^{[k]}\right)\mathbf{g}_{\tilde{\mathbf{a}}}\left(\tilde{\mathbf{a}}^{[k]}\right) \tag{54}$$

$$\tilde{\mathbf{a}}^{[k+1]} = \tilde{\mathbf{a}}^{[k]} + \delta\tilde{\mathbf{a}}^{[k]} \tag{55}$$

and then updating the path lengths $\tilde{s}[n]$ as

$$\delta\tilde{s}[n]^{[k]} = -g_{\tilde{s},n}\left(\tilde{s}[n]^{[k]}\right) / H_{\tilde{s},n}\left(\tilde{s}[n]^{[k]}\right) \tag{56}$$

$$\tilde{s}[n]^{[k+1]} = \tilde{s}[n]^{[k]} + \delta\tilde{s}[n]^{[k]} \tag{57}$$

in every iteration. In every step, weights are updated using (16) and (17) from the residuals

$$r[n] = \left\| \begin{bmatrix} \tilde{x}_V[n] \\ \tilde{y}_V[n] \end{bmatrix} - \begin{bmatrix} \tilde{x}_C(\mathbf{a}, \tilde{s}[n]) \\ \tilde{y}_C(\mathbf{a}, \tilde{s}[n]) \end{bmatrix} \right\|_2. \tag{58}$$

Figure 6 shows an example for the final polynomial curve fitted to the data points in a least-squares sense using the proposed algorithm. Points on the curve (blue circles) are closest to the data points.

Evaluation

As a demonstrative example for the capabilities of the proposed algorithm, a course including several challenging situations was utilised on which a driving tractor and trailer using path planning for steering control was simulated. As shown in Fig. 7 this includes:

- S-shaped curves to emphasize the necessity of (at least) a third-degree polynomial.
- Outlier measurements close to the desired path.
- Missing data of several meters after which the swath continues at a different angle.
- Irregular and oscillating curves that should be followed so that the swath is always within the limits given by the pickup, but not lead to oscillating steering motion.
- Sudden jumps in the lateral swath position.
- Sharp turns of the swath.

It can be seen that during the whole course the swath stays (almost always) well within the limits described by the pickup, in fact the swath is in the centre of the pickup as desired most of the time. The deviation from the pickup centre is depicted in Fig. 8a, the swath leaving the pickup range only during a very short time around 200 m of driven path length. The corresponding cumulative distribution function is shown in part (b) of the figure, showing that during over 90% of the path length the deviation is not more than 0.4 m from the pickup centre.

The algorithm has been implemented and tested in an UDOO QUAD embedded board featuring an i.MX 6 ARM Cortex-A9 1 GHz quad core CPU by NXP [15] and has been used for real-time measurement, path planning, and steering control on a tractor. With a size of 85 mm $\times$ 110 mm this hardware is small enough to be mounted together with a 2D Lidar sensor in a housing on top of the tractor cabin. In this setup, about 10 Hz update rate for 150 curve points and about 5 Hz for 200 curve points were achieved, which was shown to be sufficient for steering control at driving speeds of up to 20 km/h.

Conclusions

A robust path planning algorithm for steering control of an agricultural vehicle using a third-order polynomial representation of the curve tangent angle has been presented. Data from a laser rangefinder is used for detecting the swath to be loaded on the ground, therefore the environment provides necessary information for the mechatronic system to operate by semi-autonomously driving along the optimal path. The

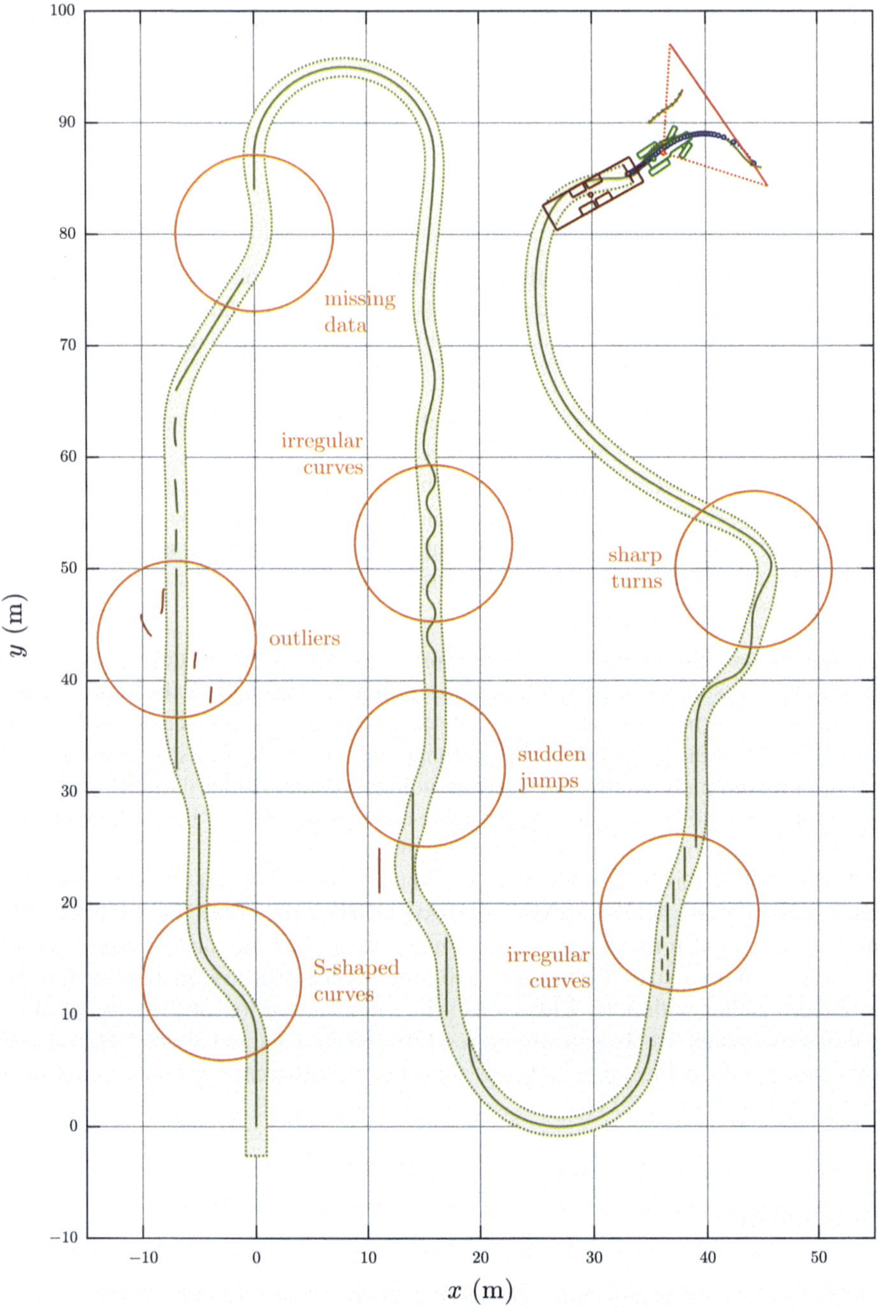

Fig. 7 Exemplary curve as driven by a tractor and trailer using the proposed path planning algorithm while following a given swath on the ground. The centre of the desired swath is shown as solid brown line, the area covered by the trailer pickup is shown in light brown with dotted edges. Undesired sources of outlier measurements are shown as dark red lines. Challenging situations are indicated by labeled orange circles

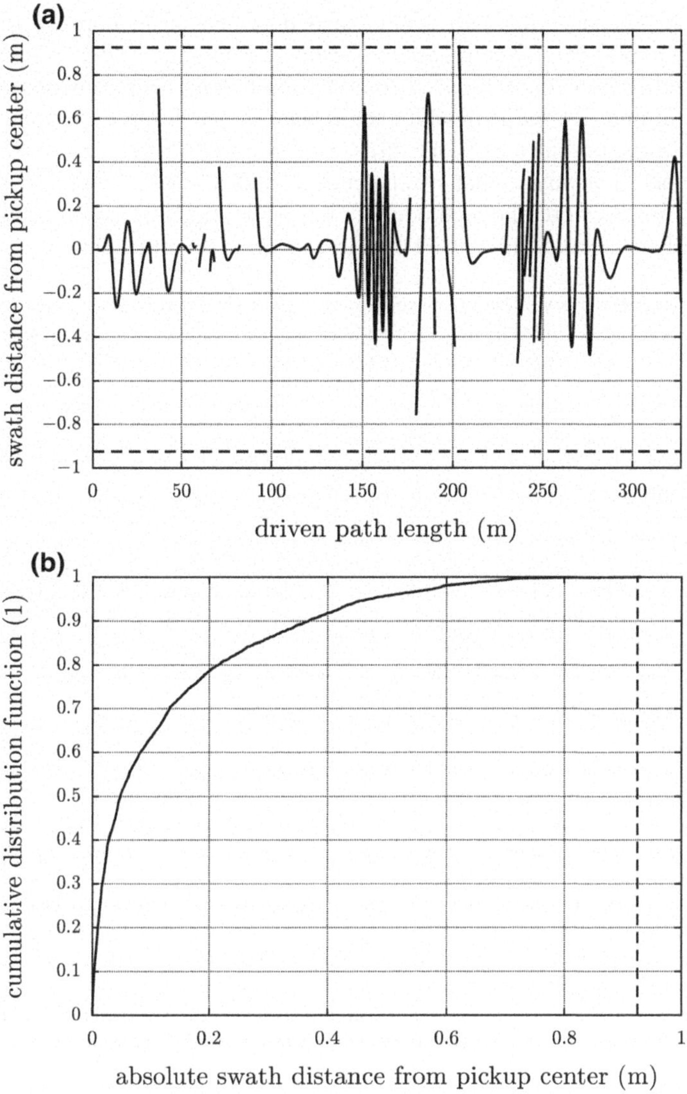

Fig. 8 Error evaluation for the pickup path from Fig. 7. **a** Swath distance from pickup center when hit by the pickup. **b** Corresponding cumulative distribution function. Pickup edges at ±0.925 m are indicated by dashed lines

optimization procedure involves a three-step initialization to improve robustness with respect to outlier measurements and ensure convergence of the following iterative optimization procedure. Robustness and stability have been demonstrated by an exemplary driving simulation in which it has been shown that even in very challenging driving situations such as sharp turns, missing data, or sudden jumps, a stable path is planned in a way so that swath centre is almost always within the width of the pickup. Furthermore, the methodology has been implemented on an embedded Linux-board and successfully tested on a tractor in real-world situations.

Acknowledgements This work has been supported by the COMET-K2 Center of the Linz Center of Mechatronics (LCM) funded by the Austrian federal government and the federal state of Upper Austria.

References

1. Gasparetto, A., Boscariol, P., Lanzutti, A., & Vidoni, R. (2015). Path planning and trajectory planning algorithms: A general overview. In G. Carbone, & F. Gomez-Bravo (Eds.), *Motion and operation planning of robotic systems* (pp. 3–27). Heidelberg: Springer.
2. Sorniotti, A., Barber, P., & De Pinto, S. (2017). Path tracking for automated driving: A tutorial on control system formulations and ongoing research. In D. Watzenig, & M. Horn (Eds.), *Automated driving* (pp. 71–140). Heidelberg: Springer.
3. Zhou, F., Song, B., & Tian, G. (2011). Bézier curve based smooth path planning for mobile robot. *Journal of Information & Computational Science, 8*(12), 2441–2450.
4. Elbanhawi, M., Simic, M., & Jazar, R. N. (2015). Continuous path smoothing for car-like robots using B-spline curves. *Journal of Intelligent & Robotic Systems, 80,* 23–56.
5. Yang, S., Wang, Z., & Zhang, H. (2017). Kinematic model based real-time path planning method with guide line for autonomous vehicle. In *Proceedings of the 36th Chinese Control Conference*, Dalian, China (pp. 990–994).
6. Choi, J., Curry, R., & Elkaim, G. (2008). Path planning based on Bézier curve for autonomous ground vehicles. In *Proceedings of the Advances in Electrical & Electronics Engineering – IAENG Special Edition of the World Congress on Engineering & Computer Science*, San Francisco (pp. 158–166).
7. Cichella, V., Kaminer, I., Walton, C., & Hovakimyan, N. (2017). Optimal motion planning for differentially flat systems using Bernstein approximation. *IEEE Control Systems Letters, 2*(1), 181–186.
8. Resende, P., & Nashashibi, F. (2010). Real-time dynamic trajectory planning for highly automated driving in highways. In *Proceedings of the 13th IEEE International Conference on Intelligent Transportation Systems*, Funchal (pp. 653–658).
9. Zhang, S., Simkani, M., & Zadeh, M. H. (2011). Automatic vehicle parallel parking design using fifth degree polynomial path planning. In *Proceedings of the 2011 IEEE Vehicular Technology Conference*, San Francisco (pp. 1–4).
10. Jiang, H., Xiao, Y., Zhang, Y., Wang, X., & Tai, H. (2013). Curve path detection of unstructured roads for the outdoor robot navigation. *Mathematical and Computer Modelling, 58,* 536–544.
11. Pichler-Scheder, M., Ritter, R., Lindinger, C., Amerstorfer, R., & Edelbauer, R. (2018). Robust online polynomial path planning for agricultural vehicles in greenland farming. In *Proceedings of the 16th Mechatronics Forum International Conference*, Strathclyde (pp. 98–105).
12. Marquardt, D. (1963). An algorithm for least-squares estimation of nonlinear parameters. *SIAM Journal of Applied Mathematics, 11*(2), 431–441.

13. Green, P. J. (1984). Iteratively reweighted least squares for maximum likelihood estimation, and some robust and resistant alternatives. *Journal of Royal Statistical Society Series B (Methodological), 46*(2), 149–192.
14. Beaton, A. E., & Tukey, J. W. (1974). The fitting of power series, meaning polynomials, illustrated on band-spectroscopic data. *Techonometrics, 16,* 147–185.
15. SECO USA Inc, All-You-Need ARM Cortex A9 SBC Development Board—UDOO @ www.udoo.org/udoo-dual-and-quad/. Accessed May 20, 2019.

The *AgriRover*: A Reinvented Mechatronic Platform from Space Robotics for Precision Farming

Xiu-Tian Yan, Alessandro Bianco, Cong Niu, Roberto Palazzetti, Gwenole Henry, Youhua Li, Wayne Tubby, Aron Kisdi, Rain Irshad, Stephen Sanders and Robin Scott

Abstract This paper presents an introduction of a novel development to a multi-functional mobile platform for agriculture applications. This is achieved through a reinvention process of a mechatronic design by spinning off space robotic technologies into terrestrial applications in the *AgriRover* project. The *AgriRover* prototype is the first of its kind in exploiting and applying space robotic technologies in precision farming. To optimize energy consumption of the mobile platform, a new dynamic total cost of transport algorithm is proposed and validated. An autonomous navigation system has been developed to enable the *AgriRover* to operate safely in unstructured farming environments. An object recognition algorithm specific to agriculture has been investigated and implemented. A novel soil sample collecting mechanism has been designed and prototyped for on-board and in situ soil quality measurement. The design of the whole system has benefited from the use of a mechatronic design process known as the *Tiv* model through which a planetary exploration rover is reinvented into the *AgriRover* for agricultural applications. The *AgriRover* system has gone through three sets of field trials in the UK and some of these results are reported.

Introduction

Space exploration has captured people's imagination since Neil Armstrong and Edwin Aldrin landed on the moon on 20th July 1969. Armstrong's first walk on the Moon demonstrated the possibilities for space exploration, and many following missions were successfully implemented. This human endeavour continues with a variety of missions such as the exploration of Mars by the *Curiosity* robotic rover.

X.-T. Yan (✉) · A. Bianco · C. Niu · R. Palazzetti · G. Henry · Y. Li
University of Strathclyde, Glasgow, UK
e-mail: x.yan@strath.ac.uk

W. Tubby · A. Kisdi · R. Irshad
RAL Space, Didcot, UK

S. Sanders · R. Scott
Veolia Nuclear Solutions, Abingdon, UK

X.-T. Yan et al. (eds.), *Reinventing Mechatronics*,
https://doi.org/10.1007/978-3-030-29131-0_5

Space robotics were initially developed for specific space missions, for instance the *Canadarm* space robotic manipulators, designed for the Space Shuttle and deployed at the International Space Station.

More recently, European stakeholders have developed the *Strategic Research Cluster in Space Robotics* to support an ambitious space robotics programme. The first batch of funding comprised five common building block projects, supported by the *PERASPERA*[1] Programme Support Activity. These include:

1. *ESROCOS*[2]: the design and development of a space robot control operating system;
2. *ERGO*[3]: a goal oriented autonomous controller;
3. *InFuse*[4]: a common data fusion framework;
4. *I3DS*[5]: an integrated 3D sensor suite;
5. *SIROM*[6]: a standard and modular interface for robotic handling of payloads.

On Earth, agriculture is a vital industry with a world population in 2015 of 7.3 billion, projected to rise to 8.5 billion by 2030. With increasing demands for better quality of food from a growing middle-class population, and a requirement for more food by the increasing population worldwide, world agriculture has entered a new era characterized by the following challenges:

- Long term demands for sustainable food production;
- Monitoring crops at the required level of resolution for better crop production;
- The urgent need for mobile monitoring platforms which can provide measurements of soil fertility in situ and in real-time for precision farming;
- A technological solution is required to meet these challenges and enable the delivery of fertilisers with minimal environmental impacts.

These requirements can be met by recent technological developments such as Earth observation, precision farming and agricultural robotics. One particular effort is through a reinvention of some of the technologies developed for space robotics that is reported in this chapter.

[1]Plan European Roadmap & Activities for Space Exploitation of Robotics & Autonomy @ www.h2020-peraspera.eu.

[2]European Space Robot Control Operating System @ www.h2020-esrocos.eu/.

[3]European Robotic Goal-Oriented Autonomous Controller (ERGO) @ www.h2020-ergo.eu/.

[4]Infusing Data Fusion in Space Robotics @ www.h2020-infuse.eu/.

[5]Integrated 3D Sensors @ http://i3ds-h2020.eu/.

[6]Standard Interface for Robotic Manipulation of Payloads in Future Space Missions@ www.h2020-sirom.eu/.

Soil Quality Measurement Challenge

The fertility level of soil is a key indicator of soil quality, and its measurement has been mostly manual until now. Due to the large areas dedicated to farming, the current practice in many countries, including for instance the UK and China, is either to:

1. Measure nutrients, such as nitrogen, potassium and phosphate, at relatively infrequent intervals, perhaps once every two years, using human operators, with typically 100 m resolution

   or

2. Measure at much lower spatial resolution through indirect soil nutrient measurement using remote sensing of the leaf area index of a plant's canopy.

A good understanding of soil fertility is critical for precision farming, as this provides the key information to map the targeted farm land, and therefore to optimize the distribution of fertiliser within a field, and to select the right type of fertiliser at the right time and at the right intervals.

Satellite data derived systems have already been used for soil quality monitoring with model-based calibration. However, in situ point measurements are still the most common technique used in agricultural, though some ground vehicles are also used on an ad hoc basis. Satellite data are then primarily used for research investigations. The feedback from end users is that both spatial and temporal resolution of satellite-based monitoring are still not adequate for precision farming. Specifically, the data acquisition rate cannot meet agricultural production demands. Nevertheless, manual measurement precision is not reliable and objective.

To make remote sensing techniques more widely available and usable for soil quality measurement, it is essential to have frequent and reliable calibration data, related to larger areas in order to support the satellite-based modelling and predictions. The current challenges for Earth observation modelling are related to a lack of this readily available calibration data.

To address the above challenges, it is believed that space robotic technologies such as the *ExoMars* or *Curiosity* rovers could offer alternative solutions, and be used as a basis for reinvention of a novel solution to revolutionise soil monitoring practice. New mechatronic systems derived or repurposed from space robotics could potentially be used for automating the acquisition of soil fertility and harvest quality data. Subsequently, this data could be used as the baseline information or ground truth for the calibration of satellite imagery data. This approach to generate real ground truth data and model driven data to cover large areas is believed to create a step change to achieve the required spatial and temporal resolution for precision farming.

Current Soil Monitoring Technologies

One basic task that can be automated on a farm is that of soil monitoring, along with and in addition to, ground coverage such as ploughing, planting and harvesting [1]. Ploughing and planting involves heavy equipment, and many farmers have indicated a preference to do it personally, because of the need to manage a field. Many mechanisms have been developed specifically for tasks such as harvesting, but the success of such systems remains at only around 66%, with fruit damage occurring on average in 5% of cases and an average cycle time of 33 s [2]. While a number of robots have been designed and deployed to enable the monitoring and treatment of crops for precision agriculture applications, many of them did not move beyond the prototype stage. Successful agricultural robots will need to meet stringent cost requirements to be affordable while performing a successful fusion of global and local reference data to navigate real fields autonomously [3]. This technical need could potentially be fulfilled through reinvention, by adapting and deploying mapping and navigation technologies available for planetary rovers in space robotics.

One of the first agricultural mobile robots was *AU-RORA*, but its sensor system restricted it to the interior of greenhouses [4]. This was a small skid-steered inspection robot that integrated a monocular camera and GPS receiver for simple automated inspection duties across an entire field as described by Bengochea-Guevara et al. [5]. The use of monocular vision simplifies the image processing, but does not support making detailed maps of tall crops or navigation of overgrown areas of farmland. Asstrand and Baerveldt [6] describe another mobile robot with monocular vision for weeding that was developed on a similar platform. Also, precise vehicular control requires wheel rotation steering and path control such as set out by Zhu et al. [7].

A Reinvented Mechatronics Mobile Platform Design

AgriRover System Requirements and Architecture

Based on analysis of soil fertility monitoring through talking to various stakeholders, including farmers and operators working in fertilisation stations both in the UK and China, the following requirements have been captured:

- Affordability for farmers;
- Intelligent behaviour;
- Robust design—Rugged design for outdoors all-weather agricultural use;
- Soil sensing capability;
- Multi-functional manipulation for monitoring, watering and fruit picking;
- Small footprint—The robot should fit in tramlines and between rows of crop (designed for Maize or Wheat) without damaging the crops;

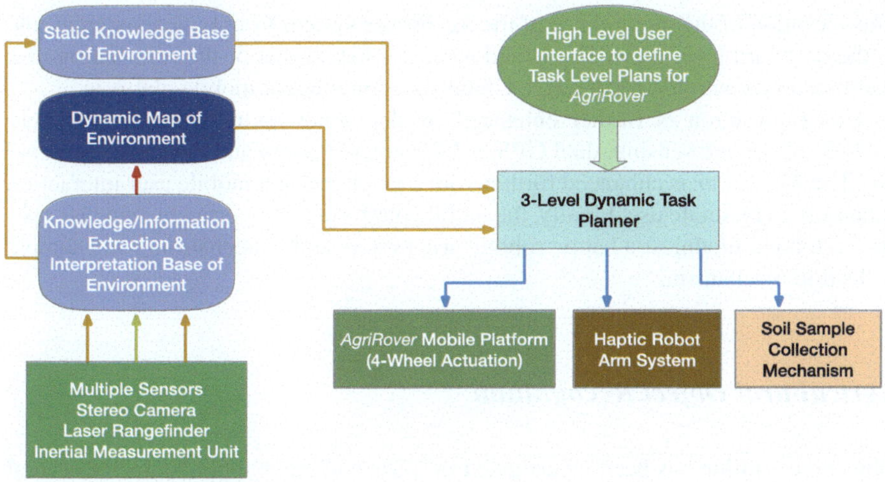

Fig. 1 *AgriRover* system architecture

- Payload capability—The platform is required to carry at least two instruments, and be able to interface with them where needed;
- Functional but aesthetically pleasing design.

Using a mechatronic design methodology referred to as the *Tiv* model [8], the research team proposed a system architecture which consists of four key hardware blocks as shown in Fig. 1. The independently driven four-wheel mechanism, the soil sensing system, navigation sensing suite and the haptic robotic arm system design are based on the analyses of space robotic systems such as *ExoMars* and *Curiosity* rovers. The wheel mechanism will ensure that the wheels can be independently controlled to cope with unstructured and challenging terrain. The haptic robotic arm system module is designed to enable a farmer to remotely control the arm so that the system can help farmers to manipulate an end-effector to harvest in a confined environment, such as humid and hot greenhouses. The soil drilling mechanism is derived from deploying space robotic working experience from partners and considering the similarity and differences between planetary soil and farming soil. A drill has been designed to capture soil at a recommended depth of 10–30 cm into the ground in order to determine the critical section of the soil for crop growth.

From these four hardware blocks, the *AgriRover* architecture is then abstracted to create the intelligent behaviour of the mechatronic system design. This is undertaken by creating a three-level dynamic task planner framework so that intelligence of the system is enabled at individual level or combination of control strategies at three levels. At the on-board computer level, the *AgriRover* can enhance its navigation and environment perception using the existing knowledge base and the dynamically changing map of the environment, by updating of the newly recognised environment. The *AgriRover* can undertake more analysis at The Cloud level, where knowledge relating to crops historical data can be accessed to enhance *AgriRover*'s behaviour.

This elevation of the intelligence of the *AgriRover* system from typical mechatronic to the on-board level on to The Cloud level is a key aspect of the reinvention of a traditional mechatronic rover into a Cloud-based intelligent mobile platform.

This reinvention is further enhanced by deploying multiple sensors, namely LIDAR, ultrasonic sensing, dual GPS, a 3-D camera system and wheel torque sensing. The *AgriRover* is enhanced further with a touch pad and mobile user interaction capability to facilitate use. Finally, the mobile platform is supported by two task specific actuation modules: a haptic robotic arm system and a mechatronic soil sample collection mechanism.

Agricultural Object Recognition

Object recognition has been investigated in space and other applications for several decades and even with the latest hardware technologies, vision-based object recognition in agriculture remains a major challenge in computer vision. Various techniques have been studied to address this challenge, these including fusion of images and other data [9], matching of image characteristic points [10], to bags of words for feature descriptors [11] and 2D image analysis to 3D point cloud processing. Among these techniques, pixel labelling allows simultaneous detection, classification and localization of objects within a 2D image.

Classification of agricultural features such as crop and plant identification have been a key challenge in automatic control, and machine learning techniques based on recognition of colours and some shapes have been developed by Yan et al. [12].

The Reinvention Process and the *AgriRover*—Sensing Mobile Platform

Progressing from the conceptual design stage of the *Tiv* model to the detailed design stage through multi-perspective modelling and simulation, a prototype *AgriRover* system has been constructed. It has two modes of operation, namely *AgriRover*-Sensing for soil nitrogen measurement and *AgriRover*-Harvester for selective harvesting. This paper focuses on the first mode of operation and its major components are shown in Fig. 2.

The *AgriRover*- Sensing mobile platform technology is inspired by space robotic research in planetary exploration, represented by the *ExoMars* rover. It addresses the specific needs of soil quality monitoring inspired by on-board chemical analyser *ChMin* on *Curiosity*. The platform technologies comprise a four-wheel mobility structure, rover control architecture, soil sample collection mechanism and a user interface for high level rover control and monitoring.

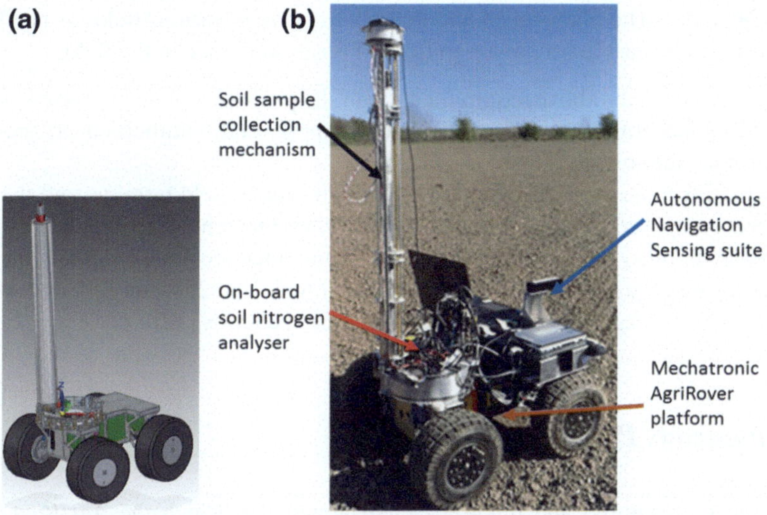

Fig. 2 One of the initial concepts **a** and **b** the final *AgriRover* platform system prototype with the main featured subsystems

A holistic and systematic approach has been taken in the process of repurposing a mechatronic system *AgriRover* inspired by space robotics. Reinvention of mechatronics in this context is defined as a holistic and systematic process of creating a new mechatronic solution for a defined application, e.g. agriculture, by exploiting and adapting the technical solutions produced in another application, e.g. space robotics. The process builds on the traditional mechatronic design process as represented by the integration of mechanical, electrical and control disciplines. In addition, the advanced computer vision-based systems, sensing systems and communication systems accessing information from the Cloud, have been deployed in order to enable the system to achieve significantly enhanced intelligence in interacting with intended objects or performing purposeful interventions within its environment.

Four Independently Driven Wheeled Platform Design

The following requirements offer guidance to the mechatronic reinvention effort:

- Long operational time—The robot is required to run for 8 h in a field from a suitable power source;
- Affordability—The platform must be affordable using off-the-shelf parts;
- Ease of Maintenance—The platform should have low maintenance requirements and be easy to repair;
- Ease of Assembly—Reconfigurable platforms are required for multiple purposes so the reconfiguration should be easy and relatively quick;

- Low Weight—The platform should be easy to move to various fields as needed. It should also be easy for a person to lift and position it once in the field.

In order to meet the above requirements, a multi-purpose mobile platform has been designed by selecting, evaluating and integrating several commercial off-the-shelf mechatronic components.

The platform has been designed to cope with typical field terrains, and the final concept is composed of four integrated wheel drives and wheel steering sub-systems for reliable mobility. It uses four in-wheel motors to reduce the weight and the size. The platform is roughly 400 mm long, 400 mm wide and 254 mm tall, and weighs about 15.5 kg (25 kg full load) with hardware and batteries [12].

Instantaneous Power Modelling

Given the unstructured and unknown environments in which the *AgriRover* will operate, it is important to model power consumption to ensure the system meets the affordability and operational time requirements. A new method for measuring the energy efficiency of a mobile platform is proposed which can be applied to any mobile technology using legs or wheels. The dynamic energy efficiency of the *AgriRover* is derived in order to establish the key energy performance characteristics of the mobile platform. This will allow for the evaluation of the instantaneous and peak performance characteristics of the mobile platform in typical soil sensing operations. In these applications, it is insufficient to measure only the static energy efficiency and as such the new approach includes time analysis.

The instantaneous power is derived from first principles. Referring to Eq. 1, the instantaneous power P is given by the average power P_{avg}, as the time interval Δt approaches zero [13]:

$$P = \lim_{\Delta t \to 0} P_{avg} = \lim_{\Delta t \to 0} \frac{\Delta W}{\Delta t} = \frac{dW}{dt} \tag{1}$$

The total power required at any instant P_T, is given by the sum of several contribution:

$$P_T = P_m + P_s + P_f + P_h + P_e \tag{2}$$

where: P_m is the power used for the displacement of the rover at any given time and is a value which differs with different traction force F and velocity V_l; P_s is the power used for steering the rover; P_f is the contribution of friction; P_h is the power associated with the efficiency and the heat loss of the propulsion system, and is proportional to the current that passes through the propulsion system. Finally, P_e is the power required for the on-board electronic equipment. Each power term can be calculated as:

$$P_m = F \cdot V_l \tag{3}$$

$$P_s = \tau \cdot \omega \tag{4}$$

The driving motors heat loss P_{hd} is a function of P_m and can be considered as the η efficiency of the rover motor and the motor driver. It is calculated as:

$$P_{hd} = P_m \cdot (1 - \eta) \tag{5}$$

The steering motors heat loss P_{hs} can be calculated as:

$$P_{hs} = P_s \cdot (1 - \eta) \tag{6}$$

When P_h will be the sum of P_{hs} and P_{hd}.

At any instant the instantaneous power P(t) can be defined as:

$$P_{(t)} = \frac{dW}{dt} \tag{7}$$

According to the total power required at any instant, the energy cost of the rover E_t can be defined by:

$$E_t = \int P \cdot dt \tag{8}$$

Merging Eqs. 2 and 7, at any given time interval from t_1 *to* t_2, the *AgriRover* energy consumption E_T is given by the integral:

$$E_T = \int_{t_1}^{t_2} (P_m + P_s + P_f + P_h + P_e) \cdot dt \tag{9}$$

Applying Eq. 9, it is proposed to define the Total Cost of Transport (TCoT) to be the ratio between the power consumption and the product of weight and velocity in real time, enhancing the static measurement introduced by Bhounsule et al. [14] when:

$$\mathrm{TCoT} = \mathrm{E_T(t)/(W(t) * V(t))} \tag{10}$$

where W and V are the weight of the *AgriRover* and the velocity at which it travels. Applying Eq. 10, it is possible to estimate the dynamic total cost of transport as shown in Fig. 3.

It is important for agricultural applications to have this measure, as it gives a more precise and continuous measurement of the cost of moving the *AgriRover* in an unstructured and unknown terrain for soil sampling. This can be used as a criterion to guide the motion planning discussed below.

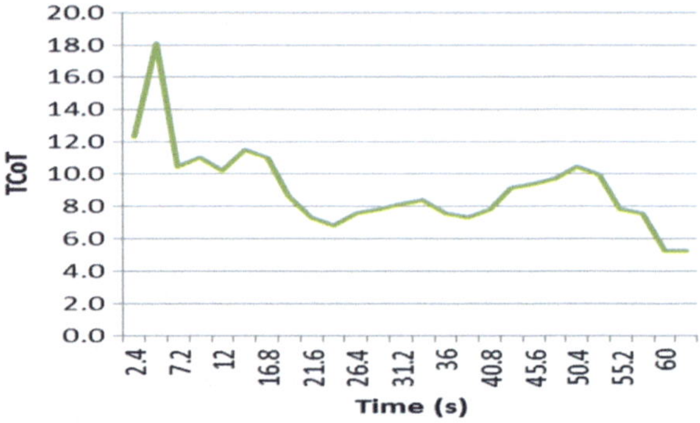

Fig. 3 A dynamic energy efficiency plot example at 30% of the maximum speed running for 18.3 m and 63 s and with an average speed of 0.3 m/s

Soil Sample Collection Mechanism

The *AgriRover*-Sensing is equipped with a laser induced breakdown spectrometer (LIBS) system for soil sensing and detection of nitrogen. The design and operation of the LIBS system in analysing a soil sample can be found in Yan et al. [15]. This paper focuses on the mechatronic aspect of the *AgriRover* system, i.e. how the sample will be collected and prepared for LIBS analysis.

Depth of soil sampling is the first issue to be addressed. For soil sampling for nitrogen measurement Miransari and Mackenzie [16] suggested a depth between 30 and 100 cm. Due to the limited payload weight of the *AgriRover*, it is important to keep the drill weight to a minimum, and deeper penetration will lead to longer and heavier sampling mechanisms. It has been therefore decided to set a depth of 50 cm based on the consultation with soil experts from the Chinese project partners. Renderings of the overall design are shown in Fig. 4.

Agricultural Object Recognition

An agricultural objects recognition software module has been developed to enhance the autonomous navigation capabilities of the rover. The module recognises agricultural landmarks such as fences, trees and buildings, taking as input the raw pixel images received by the on-board *Zed Camera*, and outputs an object membership class for each pixel.

The core pixel labelling work is executed by the Darwin framework [17], which operates as follows: (i) a preliminary classification of each pixel is computed by a machine learning algorithm, (ii) a segmentation algorithm decomposes the scene into compact regions, (iii) regions are classified according to a majority vote of the contained pixels, (iv) all the pixels in the regions are classified according to the majority vote.

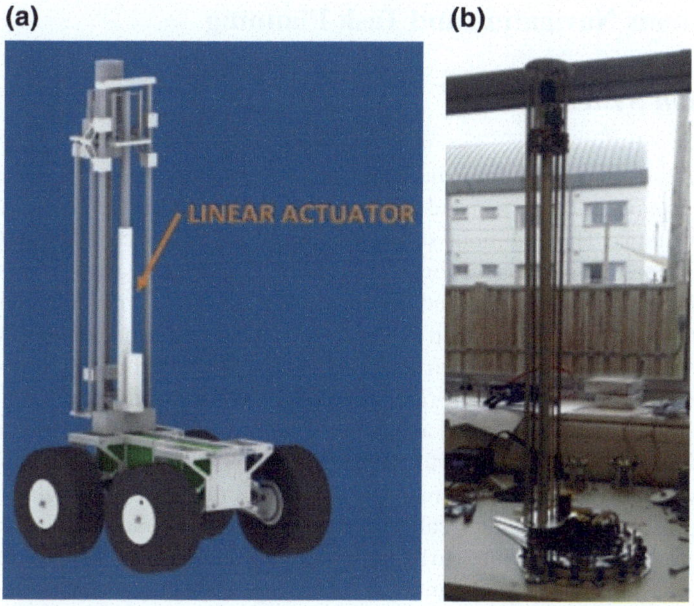

Fig. 4 Mechatronic driller design. **a** Fully assembled *AgriRover* with soil sensing capability known as *AgriRover*-Sensing with **b** a soil sampling prototype

A classification example of an agricultural object "*Cows*" is shown in Fig. 5. The left image is the original image. The right image is Darwin output. In the classification, a red colour refers to cows, the lighter blue colour refers to the sky, the darker blue colour refers to the ground, and the black colour represents unclassified areas.

Fig. 5 Agricultural object "*Cows*" recognition classification example

Autonomous Navigation and Task Planning

Navigation System Design

The *AgriRover* has been a validation platform for designing and testing an autonomous navigation system and experimenting with modern perceptual and decisional algorithms. The ambitious software architecture is depicted in Fig. 6. In this implementation, the rover is responsible for the collection and analysis of soil samples in farm fields. When a user sends the terrestrial coordinates of a point of interests, the rover reaches the locations and performs the drilling and soil analysis tasks.

The navigation system relies on seven sources of information: (i) a pair of stereo cameras, (ii) a 2D laser scanner, (iii) an inertial measurement unit, (iv) the wheel odometry sensors, (v) a GPS, (vi) a drill movement sensor, and (vii) the spectrometer for soil analysis. The rover performs its operations with two actuators, the wheel motors and the drill motors. The navigation system comprises five high level modules: (i) a SLAM module for localization and mapping, (ii) an object recognition module for obstacle detection, (iii) a planner module for path planning and movement execution, (iv) a navigation module for management of the locations of interest and (v) a drilling module for drilling and the analysis of the soil samples.

The SLAM module uses the information from the cameras, the laser, the IMU and the wheel odometry to produce a set of 3D points representing the space the rover visited, and hence to determine the current position of the robot. The map and the position are constantly updated during movement, and allow the robot to maintain perception of its environment.

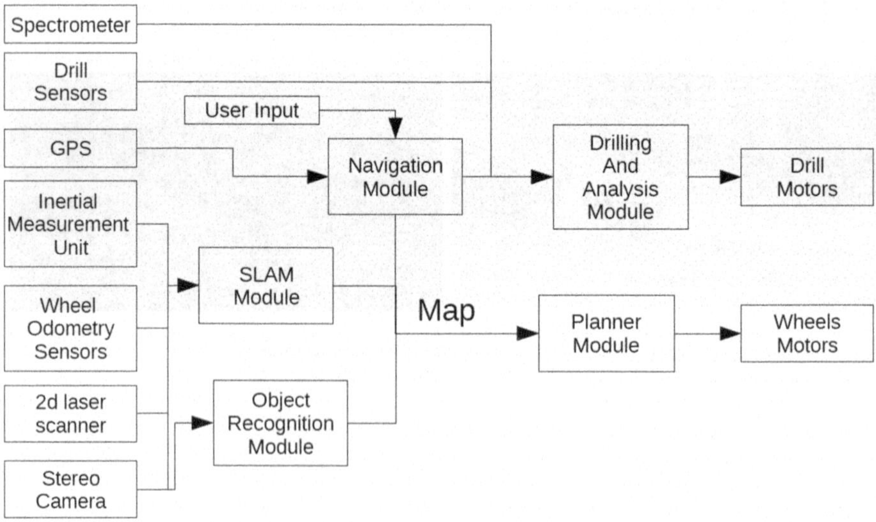

Fig. 6 *AgriRover* navigation system architecture

In the SLAM module, the implementation of RTAB-MAP in [18] available in the ROS public repository was used. The module performs a series of steps as follows: (i) a point cloud is created for each pair of stereo images by triangulation; (ii) 2D features are extracted from the stereo images and features belonging to consecutive images are compared. Their relative displacement is then used to compute the distance the rover moved in the time interval between the two image recordings; (iii) consecutive laser scans are compared to obtain another estimation of the robot movement; (iv) different estimations of the robot movement are combined by means of a Kalman filter to obtain a final estimation; (v) point clouds are overlapped at the position predicted by the movement estimations and a complete point cloud produced; (vi) if the rover re-visits a location, then a loop closure occurs, the map is adjusted and the estimated positions are updated. Thanks to the loop closure step, re-visited locations are duplicated by virtue of the estimation errors.

The obstacle detection module analyses the 2D camera images and locates obstacles within the 3D map. This module uses the public implementation of the Darwin method [17], and performs pixel classification of an image, according to a pre-trained model as described above. Once pixels are classified, they are projected onto the 3D map by means of the same triangulation performed by RTAB-MAP. In the experiment, it was necessary to pre-train a model by providing a set of manually classified images to the Darwin learning algorithm.

The navigation module keeps information about the locations the rover visited and the locations the rover needs to visit next. Whenever a new location needs to be reached, the navigation module passes the information to the planner module. Once the location is reached, the navigation module activates the drilling and soil sampling module.

The planning module computes a 2D movement path within the 2D projection of the reconstructed map, it also transforms the high level path plan into movement commands for the wheel actuators. This module uses the NAVFN [19] global path planner implementation available in the public ROS repository, in conjunction with a system specific implementation of a local path planner. It is essential for the planning stage that the rover correctly perceives its position inside the map at all time, in order to send timely correction commands to the wheel actuators if unwanted deviations are detected.

Energy Based Path Planning

The *AgriRover* should complete its soil sampling task using the minimum amount of energy. For this reason, the total cost of transport (TCoT) is evaluated as a measure of energy efficiency. Table 1 reports some values of the cost of transportation as measured during field trials with the *AgriRover*.

Table 1 shows that the speed, duration, distance travelled and average speed influence the rover's energy consumption. It is observed that the TCoT is lower either

Table 1 Field trials preliminary results

Set speed (%)	Distance travelled (m)	Time (s)	Average speed (m/s)	TCoT
15.00	37.1	58.0	0.64	4.67
30.00	18.3	62.0	0.25	9.1
50.00	20.0	43.0	0.46	7.91
100.00	18.3	22.0	0.83	6.70

when the rover is at full speed or minimum speed. This indicates that it is beneficial to select the most appropriate speed at path planning time.

In addition, a good understanding of the field's three dimensional terrain is critical for planning an appropriate path, as moving uphill and downhill along the shortest path might be more energy expensive than a longer flat path. These insights inform the 3D planner presented below.

Energy Optimization Path Planning

For a transfer task, a further distance needs to be travelled by the rover, as for instance when the rover needs to relocate to a nearby field or when the rover needs to go to a charging station. Energy can be wasted if conventional path planning methods are deployed, so an energy optimization path planning algorithm is used.

This algorithm is based on artificial potential fields with enhancement [20] with environment and rover property variables used as input. For each factor that influences the energy consumption, an artificial potential field map is calculated. All such potential field maps are then combined and the distance potential field calculated based on the terrain map. Suppose X and Y are the horizontal and vertical coordinates for a given point and X_d and Y_d the horizontal and vertical coordinates for the destination point and is gravity, then the distance potential field E_P is calculated as follows:

$$E_P = \lambda\sqrt{(X - X_d)^2 + (Y - Y_d)^2} \tag{11}$$

The potential field of height difference is calculated on the basis of the topographic map and the formula for elastic potential energy. Suppose H_P is the starting point height and H_D is the destination point height, then, a potential field for height difference E_p is calculated as follows:

$$E_{PH} = \frac{1}{2}|H_P - H_D|^2 \tag{12}$$

The two generated potential fields are combined as the weighted sum of the potential fields with each potential field having an adjustment factor based on its importance. Based on the final combined potential field, a path finding algorithm is used to find the path for the rover by simulating a free rolling ball from a high potential energy point to a lower point. An evaluation algorithm is utilized to assess how long the path is, how long the vertical component of the path is and if there is any collisions with the obstacle.

In the study, the original point is located on the top left and the destination point on the bottom right has been set and two paths have been found as shown in Fig. 7. In Fig. 7a a path found using conventional path planning and in Fig. 7b a path is identified by deploying energy optimized path planning.

The height and position relationship diagrams of these two paths can now be generated, as shown in Fig. 8. Based on this figure, the energy needed to overcome local gravitational force is calculated with equation W = m*g*h (where m is the mass of rover, the local acceleration due to gravity in ms^{-2} and *h* is the height). Here, the weight of the rover is 25 kg and the rolling resistance coefficient is 0.15. In

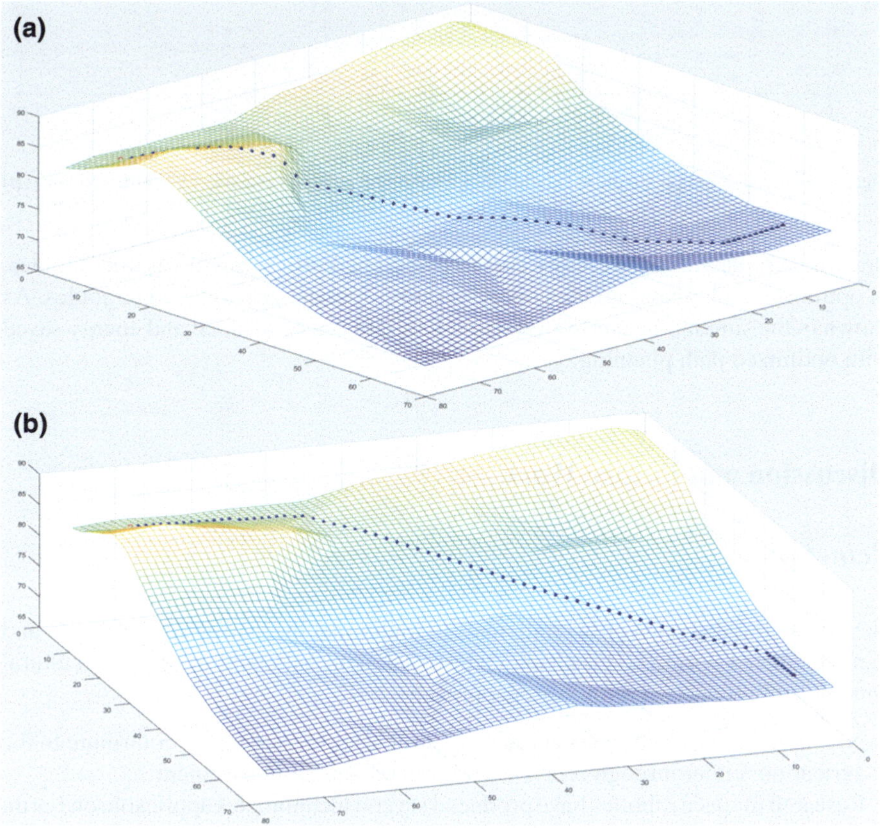

Fig. 7 Path found by **a** straight path planning and **b** the energy optimized path planning

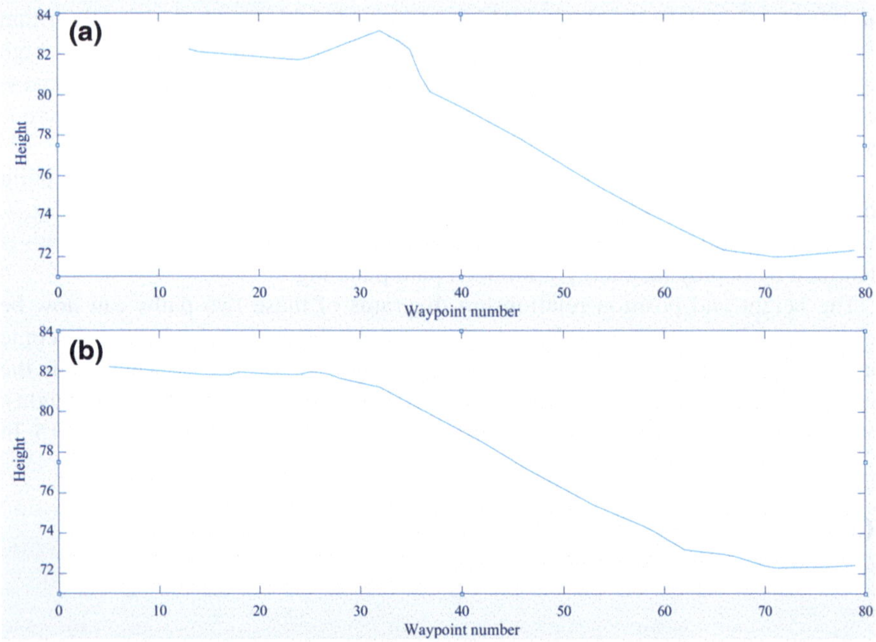

Fig. 8 Diagram of height-position relationship for straight path (**a**) and energy optimized path (**b**)

straight path planning, the energy cost to overcome Earth's gravity[7] is 406.7 joules. In optimized path planning, the energy cost to overcome gravity is 90.65 joules. As shown in the simulations, most of the uphill terrain can be avoided and energy saved with optimized path planning.

Discussion and Future Work

Reinvention

New industrial demands pose challenges to traditional mechatronic approaches and their associated systems. This chapter addressed such challenges from an agricultural context, and proposes a reinvention approach:

- New emerging technologies allow mechatronics professionals to contribute to the agriculture either through reinvention or incremental improvement.
- Research in space robotics have produced several technologies applicable on Earth, such as autonomous navigation, soil chemical composition measurement, and a

[7] $= g = 9.81\ ms^{-2}.s$

sample drilling and collection mechanism. All these technologies can offer inspiration for mechatronics reinvention.

Future Mechatronic Research Directions

This study attempts to use machine learning to train and create a capability to recognise agricultural objects. The number and types of the agricultural objects investigated within the project to date are limited and these will be extended to make the *AgriRover* system more applicable for wider crops and fields.

In order to ensure the correctness of the detected obstacles within the map and allow the navigation algorithm to compute correctly the path to the destination, a map refinement system will be added. In addition, a terrain detection module will be investigated and added to the object detection module in order to allow the robot to choose the best image sampling strategy, and the best odometry correction procedure.

Amore energy-efficient path planning and operation control system will be investigated in order to address the battery capacity challenge which is constrained by the overall weight from the design specifications.

Conclusions

The *AgriRover* system demonstrated a new way of tackling the soil quality measurement challenge faced globally through a mechatronic reinvention design process. The reinvented mechatronic solution *AgriRover* represents a holistic step forward in addressing agricultural soil measurement and management, and selective harvesting. This has shown a new technology to support sustainable, cost-effective and real-time soil quality measurement for future farming. The *AgriRover* has already generated societal impact and been defined as a key technology for a future SmartFarm project, a flagship challenge project in AgriTech collaboratively investigated by the researchers from the UK and China. It is believed by researchers that the *AgriRover* will provide farms with new technologies to enhance farm productivity, whilst maintaining environmental sustainability, by providing much needed reliable and timely soil fertility information.

Acknowledgements The *AgriRover* project is funded by the UK Space Agency under its International Partnerships in Space Programme and the authors would like to thank the Agency for its financial support. The authors would like to thank the owner of the Rushyhill Farm for its use for field trials of the *AgriRover*. Part of this work has received funding from the European Union's Horizon 2020 research and innovation programme under grant agreement No 821996 for MOSAR.

References

1. Bloss, R. (2014). Robot innovation brings to agriculture efficiency, safety, labour savings and accuracy by plowing, milking, harvesting, crop tending/picking and monitoring. *Industrial Robot, 41*(6), 493–499.
2. Bac, C. W., Henten, E. J., Hemming, J., & Edan, Y. (2014). Harvesting robots for high-value crops: State-of-the-art review and challenges ahead. *Journal of Field Robotics, 31*(6), 888–911.
3. Yaghoubi, S., Akbarzadeh, N. A., Bazargani, S. S., Bamizan, M., & Asl, M. I. (2013). Autonomous robots for agricultural tasks and farm assignment and future trends in agro robots. *nternational Journal of Mechanical and Mechatronics Engineering IJMME-IJENS, 13*(3), 1–6.
4. Mandow, A., Gomez-de Gabriel, J., Martinez, J. L., Munoz, V. F., Ollero, A., & Garcia-Cerezo, A. (1996). The autonomous mobile robot AURORA for greenhouse operation. *IEEE Robotics and Automation Magazine, 3*(4), 18–28.
5. Bengochea-Guevara, J. M., Conesa-Mun˜oz, J., Andu´jar, D., & Ribeiro, A. (2016). Merge fuzzy visual servoing and gps-based planning to obtain a proper navigation behavior for a small crop-inspection robot. *Sensors, 16*(3), 276.
6. Åstrand, B., & Baerveldt, A.-J. (2002). An agricultural mobile robot with vision-based perception for mechanical weed control. *Autonomous Robots, 13*(1), 21–35.
7. Zhu, Z.-X., Chen, J., Yoshida, T., Torisu, R., Song, Z.-H., & Mao, E.-R. (2007). Path tracking control of autonomous agricultural mobile robots. *J. Zhejiang University SCIENCE A, 8*(10), 1596–1603.
8. Melville, C., & Yan, X.-T. (2016). Tiv Model—An attempt at breaching the industry adoption barrier for new complex system design methodologies. In P. Hehenberger, D. &Bradley (Eds.), *Mechatronic futures: Challenges and solutions for mechatronic systems and their designers*. Springer.
9. Post, M., Michalec, R., Bianco, A., Yan, X., De Maio, A., Labourey, Q., Lacroix, S., Gancet, J., Govindaraj, S., Marinez-Gonazalez, X., Dominguez, R., Wehbe, B., Fabich, A., Souvannavong, F., Bissonnette, V., Smisek, M., Oumer, N.W., Triebel, R., & Marton, Z. (2018). InFuse data fusion methodology for space robotics, awareness and machine learning. In 69th International Astronautical Congress, Bremen.
10. Prince, S. (2012). *Computer vision: Models learning and inference*. Cambridge University Press.
11. Andreopoulos, A., & Tsotsos, J. K. (2013). 50 years of object recognition: Directions forward. *Computer Vision and Image Understanding, 117*(8), 827–891.
12. Yan, X.-T., Post, M.A., Bianco, A., Niu, C., Palazzetti, R., Lu, Y., Melville, C., Kisdi, A., & Tubby, W. (2018). The agrirover: a mechatronic platform from space robotics for precision farming. In *Proceeding Mechatronics 2018: Reinventing Mechatronics*, pp. 112–119.
13. Halliday, D., Resnick, R., & Walker, J. (2014). *Fundamentals of physics* (10th (Revised) Edition). Wiley.
14. Bhounsule, P., Cortell, J., & Ruina, A. (2012). Design and control of ranger: An energy-efficient, dynamic walking robot. In *Proceedings of the 15th International Conference Climbing Walking Robots Support Technol. Mobile Mach*. pp. 441–448.
15. Yan, X.-T., Donaldson, K. M., Davidson, C. M., Gao, Yichun, Hanling, Wu, Houston, A. M., et al. (2018). Effects of sample pretreatment and particle size on the determination of nitrogen in soil by portable LIBS and potential use on robotic-borne remote Martian and agricultural soil analysis systems. *RSC Advances, 8*(64), 36886–36894.
16. Miransari, M., & Mackenzie, A. (2014) Optimal n fertilization, using total and mineral n, affecting corn (zea mays l.) grain n uptake. *Journal of Plant Nutrition, 37*(2), 232–243.
17. Gould, S. (2012). DARWIN: A framework for machine learning and computer vision research and development. *Journal of Machine Learning Research, 13*(Dec), 3533–3537.
18. Labbé, M., & Michaud, F. (2018). RTAB-map as an open-source lidar and visual SLAM library for large-scale and long-term online operation. *Journal of Field Robotics, 36*(2), 416–446.

19. Hwang, H. Y. K., & Ahuja, N. (1992). A potential field approach to path planning. *IEEE Transactions on Robotics and Automation, 8*(1), 23–32.
20. Seok, S., Wang, A., Chuah, M. Y., Hyun, D. J., Lee, J., Otten, D., et al. (2015). Design principles for energy-efficient legged locomotion and implementation on the MIT cheetah robot. *IEEE/ASME Transactions on Mechatronics, 20*(3), 1117–1129.

Assistive Gait Wearable Robots—From the Laboratory to the Real Environment

Alireza Abouhossein, Uriel Martinez-Hernandez, Mohammed I. Awad, Imran Mahmood, Derya Yilmaz and Abbas A. Dehghani-Sanij

Abstract In recent decades, researchers have been able to develop intelligent assistive gait wearable robots (AGWR) capable of assisting humans in their activities of daily living (ADLs). These wearable robots have been developed to assist adults and elderly with mobility impairments, but also, to support children with motor disorders as a consequence of diseases. The reason for the rapid progress in AGWR is the advances in materials, sensor technology and computational intelligence achieved in laboratories across the globe. Unfortunately, despite the scientific and technological achievements, there exist many challenges that need to be overcome about the design, development and functionality of assistive robots. Also, there exist challenges in terms of computational intelligence methods, which are needed to make the assistive systems robust and reliable to work in indoor and outdoor environments, and on different terrains. These limitations along with lack of AGWR adaptability to the user affect the performance of wearable assistive robots, but also they reduce the acceptance, confidence and satisfaction of the individuals to wear the assistive robot on a daily basis. This chapter presents a description of wearable assistive devices, sensor technology and computational methods employed for activity recognition and robot control. Furthermore, the description of the essential parameters to achieve the user satisfaction, acceptance and usability of assistive robots is presented in this chapter.

A. Abouhossein
Shahid Beheshti University of Medical Sciences, Tehran, Iran

U. Martinez-Hernandez
University of Bath, Bath, UK

M. I. Awad
Ain Shams University, Cairo, Egypt

I. Mahmood · D. Yilmaz · A. A. Dehghani-Sanij (✉)
University of Leeds, Leeds, UK
e-mail: a.dehghani@leeds.ac.uk

X.-T. Yan et al. (eds.), *Reinventing Mechatronics*,
https://doi.org/10.1007/978-3-030-29131-0_6

Introduction

Movement dysfunction has become a major issue in ageing populations across the world as the quality of health for individuals is improving and individuals live longer [1, 2]. As big cities pollution becomes a major challenge [3], related congenital disorders influencing somatosensory and motor control disorders in the newly born children have led to the mobility challenges [4].

To overcome movement challenges specifically to provide a functional level of gait in both of these categories as well as other groups of people with mobility problems, assistive gait robotic devices have become the first line of choice in rehabilitation [5]. Human locomotion is accomplished in an efficient manner by the complex interaction between somatosensory [6, 7] and motor control [8] along with accurate coordination between multiple joints and upper/lower extremities.

A key feature of human gait is its ability to conform well to changing terrains and conditions underfoot. However, the most Assistive Gait Wearable Robots (AGWR) currently designed in the market lack this adaptability, which forces the user to learn and become adapted to the designed system. Understanding the gait parameters of those whose mobility is impaired is essential in the design and performance evaluation of AGWR devices.

The AGWRs most commonly addressing the mobility challenges are exoskeletons which work in parallel with the human body (human coupling idea) [9] to improve or enhance walking, orthotics which help to rehabilitate patient's impairment [10] or prosthetic systems which facilities restoration of the human walking impairment. There are currently several different types of systems available to capture and register the gait parameters to help to objectively design or evaluate AGWR [11]. Commonly available systems are photogrammetry (video-based system) [12, 13], wearable sensors [14, 15], and inertial measurement units (IMUs) [16, 17].

In the design and evaluation of the many robotic assistive devices, it is a common practice to use a video-based motion capture system which requires a large laboratory space, major preparation and lengthy patient debriefing before starting to gather the motion data. In addition, the price tag of a motion capture system is a major burden for biomedically related research groups in higher education. The intent of portable sensory technologies, along with sophisticated machine learning methods, has set a new height in front of mobility researchers to use wearable sensors for finding insight into design, performance evaluation and control of assistive robots [10]. These findings will facilitate the use of AGWRs in moving from the laboratory to the real-environment for the purpose of rehabilitation. The chapter is organized as follows:

- A brief review of major exoskeletons available for adults and children is considered with an eye on how they may provide adaptive assistance to the individuals with the gait impairments based on their control loops, metabolic energy consumption and gait parameters.
- The role of gait parameters in the design and control of assistive robots.
- Performance indices for gait evaluation.
- Conclusions.

Review of Exoskeletons for Human Assistance and Rehabilitation

Robotic exoskeletons are assistive wearable devices (AWDs) that can be classified into assistive, rehabilitative and ability enhancing [18] and are divided into four sub-groups: lower extremity exoskeletons, upper extremity exoskeletons, specific joint support exoskeletons, and full body exoskeletons. In this section, a brief review of exoskeleton devices is provided based on gait parameters features expressed in relationship to the control and metabolic energy consumption.

The ReWalk exoskeleton [19, 20] is designed by ReWalk Robotics Company in the US and is a well-known example of an assistive robotic exoskeleton for adults (Fig. 1a). ReWalk is proposed to be used for the gait rehabilitation of those patients with a complete Spinal Cord Injury (SCI), and hence no muscle activity or control in their lower limbs. One advantage of the ReWalk exoskeleton is that the operator can alter the knee and the hip angles to implement a of motion training for the paraplegic patients [19].

The Vanderbilt exoskeleton [21–24], shown in Fig. 1b, was developed by Vanderbilt University for paralyzed individuals to help them to accomplish some daily activities including walking, sitting, and climbing stairs. There are two different products: Indego Personal and Indego Therapy. The Vanderbilt-Indego exoskeleton works in combination with Functional Electrical Stimulation (FES) to contract and relax the paralyzed muscles of the user. The idea is to combine metabolic and robotic power sources while obtaining both the physiological advantages of FES and the control advantages of the assistive robotic exoskeleton. Therefore, the control structure includes two different control loops: a motor control loop and a muscle control loop which are essential for an assistive device that is possible to become adapted to the patient's needs.

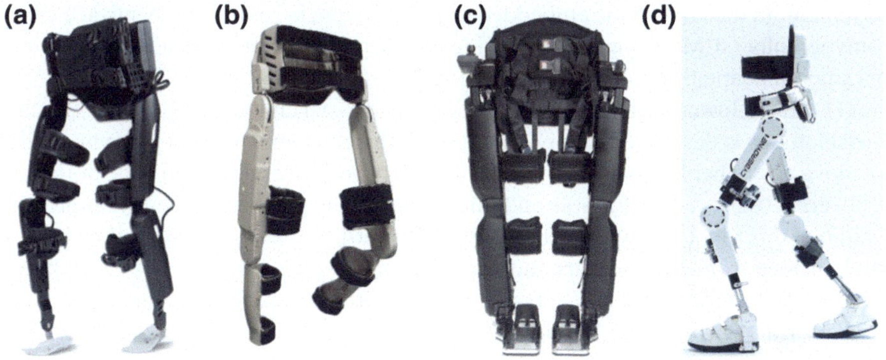

Fig. 1 Lower limb exoskeletons for assistance to walking: **a** ReWalk, **b** Vanderbilt, **c** REX, **d** HAL

In addition, Indego has a fall detection system, which allows the prediction of a potential fall. This predictive capability allows the exoskeleton to make quick adjustments to a user's position to minimize any risk of injury while in case of any power failure, knee joints are locked, and hips joints are free. The Vanderbilt team is investigating to integrate methods for postural balance and stability of the user into this exoskeleton.

REX [25, 26] is a lower limb exoskeleton from REX Bionics Ltd, which can assist subjects with many different mobility disorders, such as different levels of spinal cord injury. This exoskeleton, shown in Fig. 1c, has 10 DoFs that assist in movements for flexion/extension and abduction/adduction at hip joints, flexion/extension at knee joints, and at the ankle joints dorsiflexion/plantarflexion and eversion/inversion. REX is controlled by a joystick and can be used for many different injuries. Thus, the robot is not capable to recognize the human movement, and requires the human to be capable of using a joystick. In other words, REX is only suitable for manual wheelchair users who have control of their upper-body. Using the joystick, the robot allows the user to walk, sit, stand, turn, and navigate stairs and slopes.

The Hybrid Assistive Limb (HAL) [27] is one of the most interesting of robotic exoskeletons (Fig. 1d). It was developed by the University of Tsukuba in Tsukuba, Japan, and cybernics technology is used in the exoskeleton, which makes HAL a unique design. Cybernics is a multidisciplinary area which combines neuroscience, mechatronics, and information technologies. In other words, brain signals are used in the control loop to determine the intention to move any limb. The assistive device, HAL-1 [27], was the first developed prototype of HAL in 1999, and was only a lower-body exoskeleton. The focus was on assisting immobilized subjects to perform Activities of Daily Living (ADLs). The leg structure of the full-body HAL-5 exoskeleton powers the flexion/extension joints at the hip and knee via a DC motor with a harmonic drive placed directly on the joints. The multi-sensory system is embedded in a special shoe, thigh and waist belt of HAL, a mutual connection between the user and the HAL system, to minimize the joint resistance and boost movement. In the HAL-5 the embedded sensory system is coupled with surface electromyography (sEMG) sensors to provide the user with a better prediction resulting in a smooth adaptation to the user during ADLs.

ATLAS is a lower limb exoskeleton designed to assist children [28] (Fig. 2a). This exoskeleton was designed to assist the sagittal plane movements of a girl who could not move any of her limbs. This robot, built with 6 DoFs, was designed to be portable, lightweight (6.5 kg), comfortable and safe for providing gait assistance. An improved version of this robot, known as ATLAS 2020 [29, 30], was designed to easily adapt to children aged from 3 to 14 years and to provide them with assistance in 3D walking. An interesting feature of this robot is its capability for balance control using an auxiliary frame. This robot has been recently upgraded to include the capability for self-balance control.

Wearable Ankle Knee Exoskeleton (WAKE-up) [31, 32] is a further multi-joint lower limb exoskeleton designed for rehabilitation of children with neuromuscular diseases. WAKE-up is not a full lower-body exoskeleton, as shown in Fig. 2b. It is a modular exoskeleton involving two separate joint modules: a knee joint module and

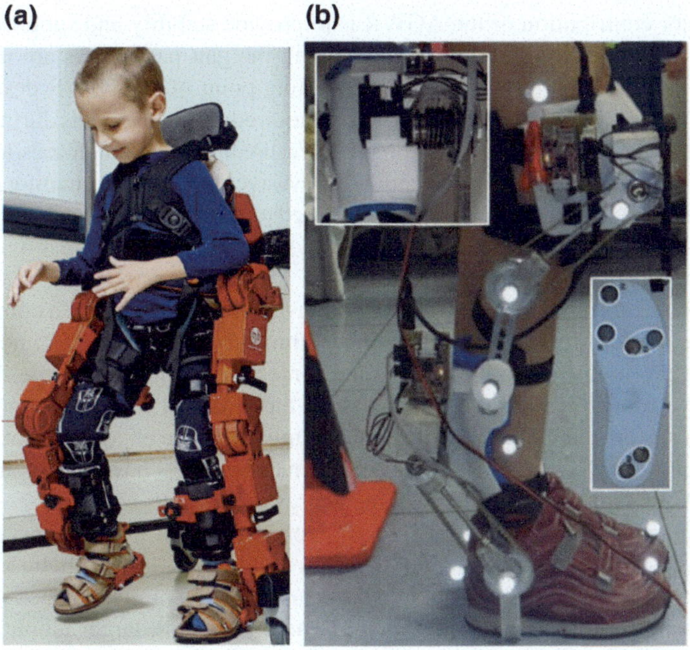

Fig. 2 Lower limb exoskeletons to assist children with gait impairments. **a** ATLAS, **b** WAKE-up

an ankle joint module. The WAKE-up could assist sagittal plane movements only, and the target age group is from 5 years old to 13 years old children with neuromuscular disease such as Cerebral Palsy or foot drop (CP). In WAKE-up exoskeleton the knee and the ankle stiffness can be adjusted via a software by an expert operator. However, such features cannot be adapted to the needs of the user automatically by the system.

Assessment of Gait Parameters in Design and Performance Evaluation and Relation to the Assistive Gait Wearable Robots (AGWR)

As discussed in the introduction, human gait itself is a functional process which requires numerous physiological systems to work congruently with the central nervous system (CNS) to achieve ambulation with minimal use of metabolic energy expenditure [33, 34]. Normal walking with a feasible physiological energy expenditure requires dynamic and static stability to provide body support during stance and swing phases to control progression of the center of mass (CoM) in a smooth and regulated manner.

The major contribution of the AGWR is to provide stability and support that the user lacks in ambulation, therefore understating the gait parameters are important in design and assessment of AGWR. The starting point of an AGWR design is an estimation of the maximal and minimal forces required across the joints during certain postural balance. It is challenging to directly measure the segmental forces/torques in the knee joint as it requires the utilization of invasive methods. Estimating dynamic loads are far more difficult in an AGWR compared to static loads that can be estimated at certain postures, and this is especially true in systems with multiple moving parts, e.g. hydraulic/pneumatic systems [34].

To overcome the challenges involved with invasive measurements, a combination of motion capture (Mo-Cap) and multi-body simulation [35, 36] is commonly used to estimate the joint maximal and minimal forces and moments for the design of any assistive robots including implants. Once the maximal forces and torques are obtained, a system can be designed which can minimize the energy cost of ambulation. It is important to mention here that experimental studies have shown ankle-foot orthotics(AFO) can minimize the walking energy costs [35].

As adaptability of available assistive devices has not yet been achieved despite technological advances, in unilateral transfemoral amputees the prosthetic leg cannot have the similar metabolic energy expenditure as a healthy human leg does [36]. On the other hand, in unilateral lower extremity amputees with larger amputation the ability of adaptation decreases. An example is transfemoral amputees who have a higher prevalence of circumduction and/or vaulting [37] (Fig. 3) than transtibial

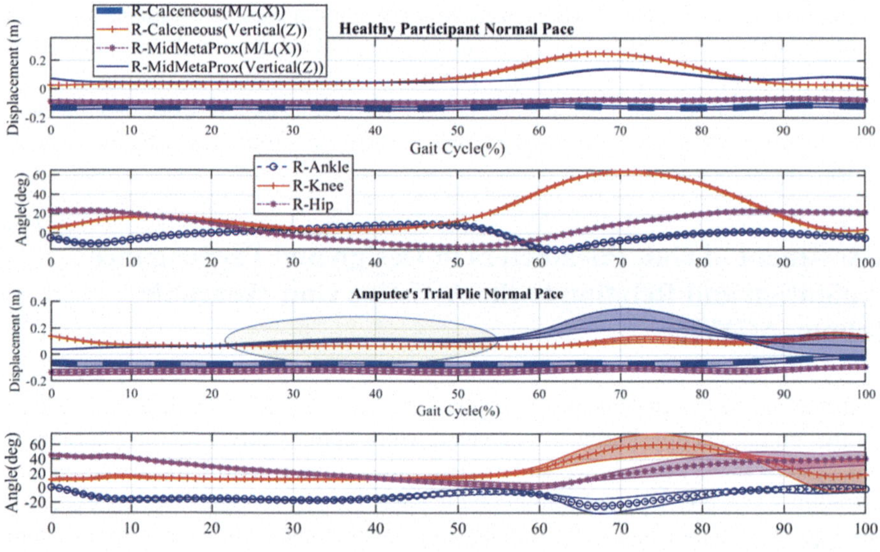

Fig. 3 Vaulting of a transfemoral amputee in comparison to the non-amputated participants. The intact ankle must flex and rise to produce the needed clearance for the amputated side (Black ellipse shows the rise of the passive calcaneus marker from the ground on the intact side of a transfemoral amputee)

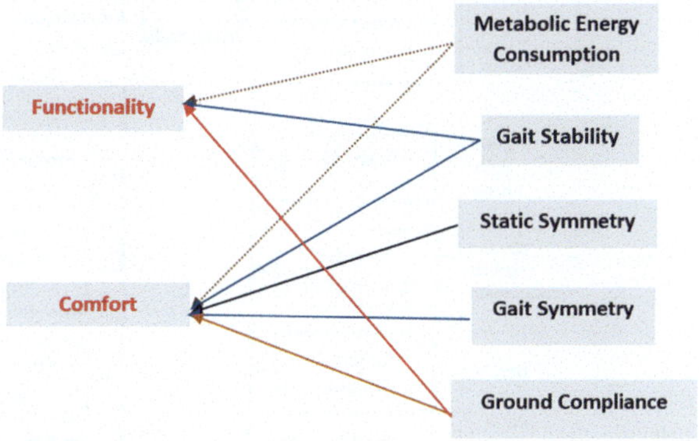

Fig. 4 Relationship between functionality and comfort of a prosthetic knee and gait parameters of amputees

amputees. As it is shown in Fig. 3 such gait abnormality can be identified using video-based Mo-Cap system vertical displacement of the amputees' ipsilateral leg (Fig. 3).

The level of adaptability of a user to an AGWR may suggest their comfort and confidence to the system. It has been a difficult task to identify what parameters affect both the functionality and the comfort of a lower limb prosthesis, and how to evaluate them to improve the overall amputee mobility. It is of significant interest to be able to ergonomically design an adaptive AGWR system that is comfortable and achieves user confidence for the desired gait components [38].

The quantitative assessment of the assistive gait wearable devices and technologies is crucial for their performance evaluation. To evaluate the wearable devices such as exoskeletons, orthoses, and lower limb prostheses, different performance indices are used based on biomechanical and physiological parameters. These parameters are utilized directly or indirectly to evaluate the overall system functionality and the level of user comfort and reduced mental efforts due to the human-robot interaction. Figure 4 shows the relationship between comfort and confidence as subjective parameters for a user of assistive devices which are linked to objective measures that are commonly identified in the laboratory using a Mo-Cap system.

As suggested, the objective is to design an ergonomical AGWR which is continuously updating its subjective parameters by evaluating the objective parameters of the user's gait to suggest an adaptive solution. The human-robot interaction and functional outcomes of AGWR are mutually related and affect one another. The impact on the user comfort while wearing the AGWR can be further classified and evaluated with the parameters shown in Fig. 5.

On the other hand, the other intention of this chapter is to suggest how one can use the on board embedded and wearable sensors to simply transfer the gait data from Mo-Cap laboratory to the real or perhaps clinical environments. The following section describes the performance indices and parameters shown in Figs. 4 and 5.

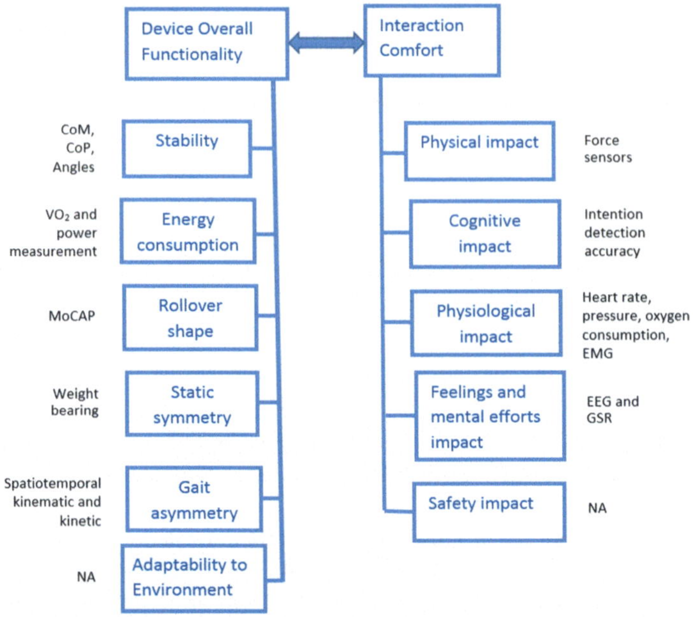

Fig. 5 The parameters used for assessment of assistive gait wearable robotic systems functionality and comfort level

Methods and Challenges for the Data Collection, Analysis and Evaluation of Gait Performance

To evaluate performance and level of adaptability of the wearable devices such as exoskeletons, orthoses, and lower limb prostheses, different performance indices are used based on biomechanical and physiological parameters. These parameters are utilized directly or indirectly to evaluate the overall system functionality and the level of user comfort and reduced mental efforts required due to the human-robot interaction. The human-robot interaction and functional outcomes of AGWR are mutually related and affect one another as mentioned before. In the following section the basic method of motion capture laboratory-based data collection is discussed. This is explained based on studies at the University of Leeds. A brief description of the parameters and relationship of these to the objective parameters obtained via photogrammetry is provided here as well.

Photogrammetry

In the photogrammetric approach, infrared optical cameras are used as a vision system, and fiducial is placed on the participant's anatomical site in the form of passive or active markers. These components are used for measurement of kinematics and spatiotemporal parameters of human gait, while force platforms are used for directly measuring ground reaction forces to estimate the gait dynamic parameters.

The photogrammetry approach is very expensive, time consuming and limited to be used inside the laboratory spaces and is not suitable for outdoor activities. In the University of Leeds, a Qualisys Motion capture system[1] composed of 12 Qualisys Oqus, 4 cameras and 1 Qualisys Oqus 310 high speed video camera, was used for the photogrammetry approach and understanding the objective measures required to link to the adaptability of the users to subjective parameters of comfort and confidence. The camera system was configured to collect data at 400 Frame per second (fps) through QTM software.

In addition, AMTI force platforms[2] embedded in the floor were used to measure gait events and ground reaction forces (GRFs). The force plate data was set to 1200 samples per second. Another portable force platform was used to collect GRFs during ramp and stair activities. The preparation time for every activity starts with calibration of the required area, placing reflective markers on the anatomical sites to establish the location of the joint centres and define limb segments tracing the location and their orientation in the three-dimensional laboratory environment. The results of the three-dimensional gait and motion analysis are highly dependent on the placement of these markers, which is one of the major sources of error in kinematic data. Figure 6 shows an above-knee (AK) amputee performing some activities of daily living in the laboratory environment.

The impact on user comfort while wearing the AGWR can be classified and evaluated with the following parameters:

Physical impact—Results from the physical contact forces (normal and shear forces), between the wearer and the device are critical for the user's comfort. These parameters can be measured indoor using force platforms and gait analysis tools through an inverse dynamics calculation, in the outdoor setup, such measurement can be metered with force wearable sensors such as a force-sensing resistor (FSR) and Flexiforce sensors. However, accurate shear force measurement is still a challenge using wearable sensors and further research in this area is required.

Fig. 6 Sensor data collection with the motion capture system while a subject performs some activities of daily living in a laboratory environment

[1] Qualisys AB, Sweden.

[2] Watertown, MA, USA.

Cognitive impact—Results from the likely difference between the intention of movement and the actual movement of the wearable device. This requires more accurate intuitive control system based on activity recognition and machine learning approaches. The intention detection accuracy, reliability, and processing speed of this intuitive control should be considered as well for interaction performance index and relating the questionnaire based subjective measures to more of objective indices which may include the physiological signals.

Physiological impact—This is measured by estimating the heart rate, blood pressure, oxygen consumption and muscle activity to assess the effects on the user due to wearing AGWR.

Feelings and mental efforts impact—The feelings and mental efforts of the wearer when they use the device has important impact on evaluating the performance of AGWR. Current approaches to assessing these parameters are based on subjective methods using questionnaires filled in by the users. To have some measurable objectives it is suggested to link the effect of mental efforts to the parameters obtained in Electroencephalography (EEG) and/or Galvanic Skin Response (GSR) sensors or monitoring the facial expressions.

Safety impact—This is related to the functional and biomechanical safety of the user and the device interaction while using it in different environments and terrains. The safety criteria should be carefully considered while designing any AGWR system. Although ISO 13482 standard sets the safety requirements and criteria of assistive robots, including wearable exoskeletons, it does not provide considerations for the measurement of biomechanical safety. The functional safety can be related to the user-device stability, maintaining postural balance or the device fail-safe criteria, while the biomechanical safety will be related to the forces transmitted to the wearer's joints and spine which may create serious problems and damage leading to the permanent impairment. The biomechanical safety parameters can be estimated from the kinematics, kinetics and gait analysis tools used in the motion capture laboratory, which are limited to indoor use.

Among these, physical impact can be measured directly or indirectly using force and kinetics sensors. The impact on cognitive, physiological, feelings and mental efforts, and safety aspects, estimated using indirect or computational analysis, may affect the use of assistive robots in real-time. The physical impact is easier to be used in the real-time assessment, evaluation, and control. Future technologies may provide more accurate and fast estimation of cognitive, physiological, feelings and mental efforts, and safety impacts which can affect the development of new generations of smart assistive rehabilitation devices and robots. The overall performance of the wearable system functionality can be evaluated based on the following parameters:

Stability (static and dynamic balance)—Gait stability is often described as the ability to recover from perturbations, which arise during performing activities of daily living (ADLs) from internal sources such as neuromuscular and external sources (e.g. wind, surface friction and/or uneven surfaces) [39]. This means that the individual postural stability depends on the neuro-musculoskeletal capacity and the type and magnitude of external perturbations an individual encounter during any ADLs. Hence, good balance and stable mobility involve continuous control and regulation of the human joints and muscles through a reactive CNS to determine the body's posture in relation to the environment and maintain the body's center of mass (CoM) within the base of support.

The stability is classified into static and dynamic stability, which are needed for quiet standing and to maintain postural balance during any movement. A study on stability [40] has shown that restriction on the ankle-foot joints will reduce dynamic stability; specifically stability is more jeopardized during swing phases than stance phases. This provides the explanation that CNS is such an important element to coordinate well among the joints and little neurological impairment leading to the joint dysfunction may result in a large instability.

Two systematic approaches commonly used to evaluate the stability are the Lyapunov exponent of human gait, and the interaction between the Centre of Pressure (CoP) and Centre of Mass (CoM). The former objective index can use kinematics data obtained using motion capture systems or wearable sensors such as accelerometers. The latter, can be evaluated using CoP and CoM which in turn can only be measured using gait laboratory-based analysis specifically the force platforms.

Energy Consumption—Energy consumption plays a significant role in evaluating the device performance. There are two types of energy associated with wearable assistive robots. The first type is the metabolic energy consumption of the user when s/he interacts with the device. The second type of energy consumption is associated with the energy consumption and efficiency of the device. Efficient device interaction with a human user should encourage less overall energy demands by the user. In users, metabolic energy consumption has been measured using VO2max, while the assistive device energy consumption has been measured using mechanical and electrical power formulation and physical measurement obtained from embedded sensors.

Rollover shape—The knee-ankle-foot (KAF) roll-over shape (ROS) is a scientific method which has been used to compare performance and design of different prosthetic foot types. At the University of Leeds, the influence of the prosthetic components (i.e. knee and foot) on the knee-ankle-foot roll-over shape in a unilateral transfemoral amputee was evaluated (Fig. 7). The kinematics of the centre of pressure (CoP), lateral knee, and ankle markers path were collected and processed to obtain ROS. The results were used to fit a circular shape arc to obtain the radius of curvature (ROC). This rollover shape is measured using gait analysis and motion capture systems in the laboratory and it cannot currently be measured using wearable sensors.

Static symmetry—Static symmetry refers to the symmetry in weight-bearing load between left and right legs. This symmetry represents the user confidence while using the assistive device, which highly depends on the device design, functionality and user her/himself including the level of muscle symmetry across the joints. This can be measured using wearable force sensors inside the insole of any AGWRs.

Gait symmetry—Gait symmetry is one of the assessment criteria used to evaluate the performance of the wearable assistive devices and robots. It refers to the percentage of the similarity between the left and right legs. The level of symmetry in human's body is related to muscle level of activation and control of CNS over the

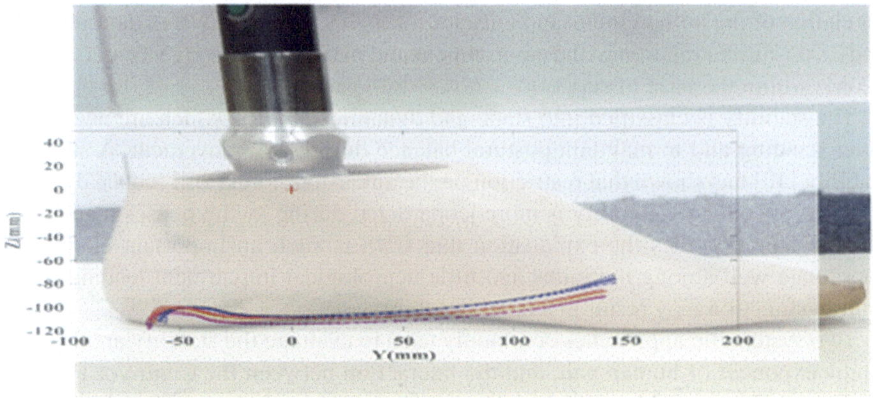

Fig. 7 A typical rollover shape of amputee with Rheo3 knee and Venture College Park prosthetic foot

muscle and the joint. It is defined as a performance index of abnormal walking and used for clinical assessment [41, 42]. The gait asymmetry can be evaluated using either spatiotemporal such as stride length, stance time ratio, step time, gait events or kinematics such as angles and velocity and/or kinetics parameters such as joint forces and ground reaction forces (GRFs) of the gait. This can be measured using both, motion capture systems or wearable sensors [43].

Ground compliance (adaptability to the environment)—Adaptability to different terrains and environments is one of the challenges which needs to be presented to the AGWR functionalities, design and control. Humans have great ability to conform their joint kinematically redundant to adapt to different terrains. The AGWR should be smart enough to be able to adapt and comply with different terrains and environment as an ordinary and healthy human does. This adaptability depends on the reaction of the mechanical system design and the control systems to the altered environments. This can be measured indirectly using wearable sensors or using motion capture system by monitoring the change in kinematics, kinetics of the device and the user compensation.

Challenges in collecting video motion capture system—As discussed thus far many of the mentioned parameters can be measured using video-based motion capture system. However, there are several challenges in collecting the video motion capture data, which must be kept in mind when such a process is considered. The setup of the video-based motion capture is itself a lengthy process as calibration of the environment is necessary before starting. The process of subject preparation and debriefing is lengthy and requires palpating the proper location of the participant's body which may introduce error into the process of the data collection. Another challenging aspect of this form of recording is the interference of light or the participant's body parts, the reflective markers may be blocked causing data loss. In the following section we briefly explain what and how wearable systems were used to estimate some of the parameters that have already been discussed.

Wearable Measurement Systems

Wearable sensors play an important role in the development of assistive gait wearable robots due to their lightweight, small size, and low energy consumption offered. These sensors, together with machine learning methods, open up opportunities for the development of robots capable of recognizing human movements, and control of stability and balance. Commonly, angular velocity, acceleration and muscle activity are obtained using wearable sensors. Additionally, the successful development of the next generation of exoskeletons, composed of soft and lightweight materials, need to integrate wearable measurement systems for analysis of kinematic data during activities of daily living (ADLs). Kinematic data has been employed for activity recognition purposes to provide safer assistance to humans while performing ADLs. The following paragraphs give an example of the use of wearable sensors and computational methods for activity recognition.

Wearable sensors for walking activities—Inertial measurement units (IMUs) were used for data collection and recognition of locomotion activities (Fig. 8). Multiple IMUs, attached to the lower limbs of subjects were used for collection of angular velocity data from level-ground walking, ramp ascent, and ramp descent activities. These data, together with machine learning methods, allowed the recognition of walking activities, gait phases and gait events in a laboratory environment (Fig. 9). Computational methods based on Artificial Neural Networks, Bayesian Networks and Support Vector Machines have shown to be highly accurate for recognition processes in assistive devices [44, 45]. It has been argued that high recognition accuracy is related to the controlled laboratory environment, and the recognition process could be affected in outdoor environments [46]. Findings suggest that location of placing the system on human body may introduce more errors than the type of outdoor environment.

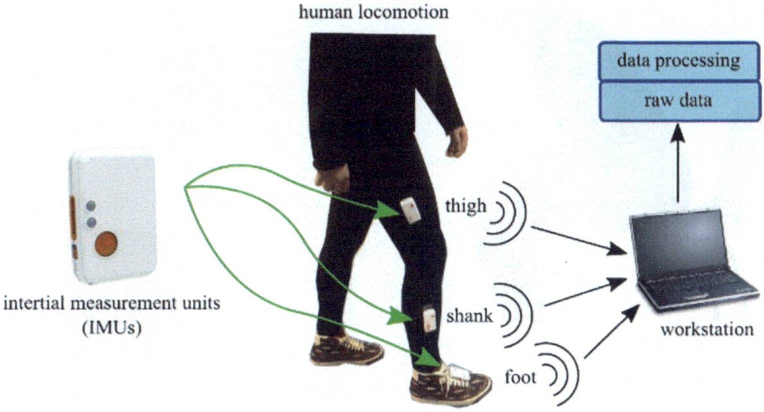

Fig. 8 Wearable sensors employed for collection of kinematic data and recognition of walking activities

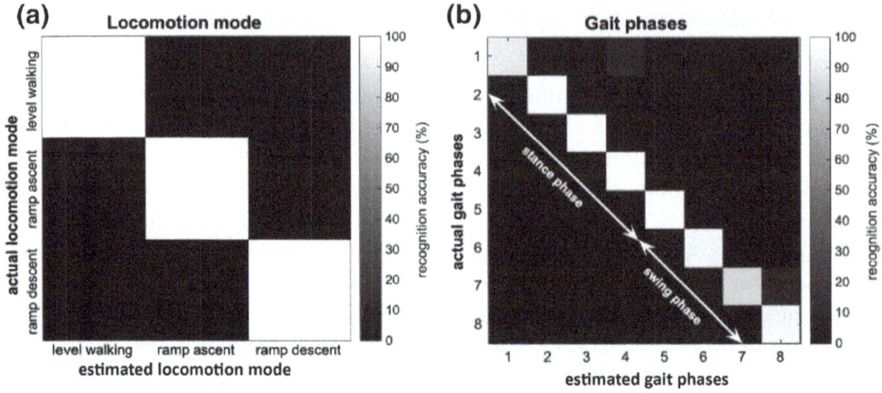

Fig. 9 Human activity recognition using wearable sensors. **a** Recognition of walking activities and **b** Recognition of gait phases and events

Wearable sensor for sit-to-stand activity—The recognition of sit-to-stand (SiSt) and stand-to-sit (StSi) activities, has also been studied in laboratory environments using wearable sensors. In the experiment shown in Fig. 10, the participant was asked to perform multiple repetitions of SiSt activity to collect acceleration data from one IMU. These data were processed by a probabilistic approach for recognition of sit, transition and stand states during the SiSt activity [47]. In addition, multiple segments that compose the transition state were recognised, which is important to achieve a robust and accurate control of assistive robots (Fig. 11). Other aspects that need to be considered in a real or outdoor environment could affect the recognition accuracy

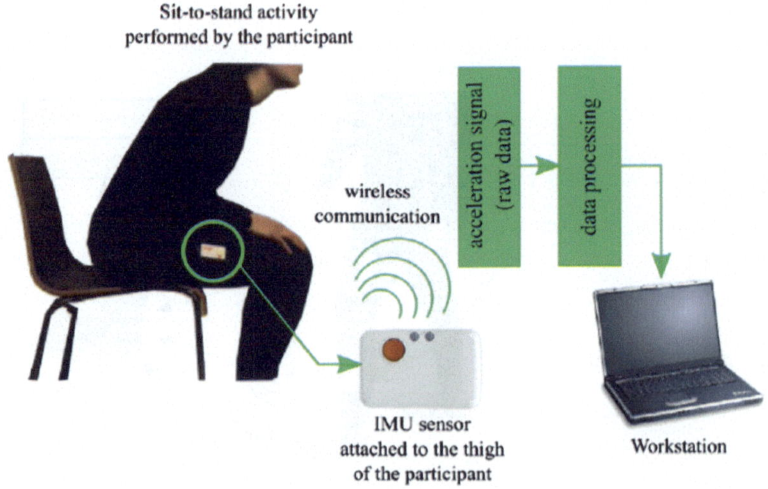

Fig. 10 Inertial measurement units employed for recognition of sit-to-stand and stand-to-sit activities

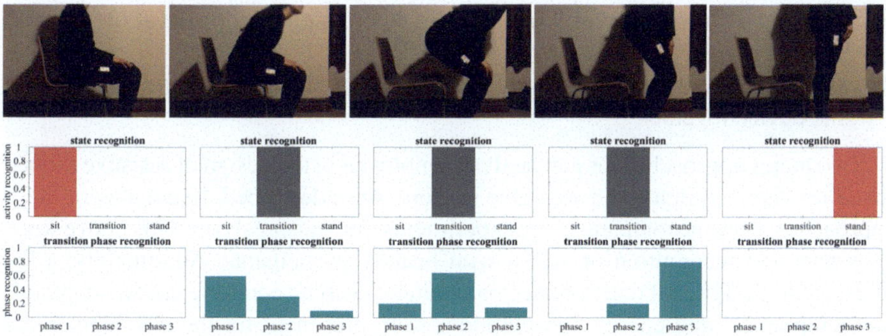

Fig. 11 Results from the recognition of sit-to-stand and stand-to-sit (top) Participant performing the activity (middle) Recognition of sit, transition and stand states (bottom) Recognition of events during the transition state

of SiSt. Some of these aspects are the delays in the pre-processing steps to smooth the signal, type of chair, optimal location of sensors and speed to move from sit to stand.

Challenges with wearable measurement systems—Wearable sensors have been successfully used for recognition of human activities. However, these results have been achieved under well-controlled laboratory environments. To achieve a similar performance in real or outdoor environments, there are various factors that need to be addressed in software and hardware. For instance, delays in recognition response due to pre-processing steps, noisy measurements, optimal location of wearable sensors on the human body, calibration of sensors over time, real-time software, context awareness, batteries and unexpected activities. Even though some of these aspects are already under investigation by researchers across the world and promising outcomes are emerging, other aspects remain a challenge.

Conclusions

The healthcare related market of assistive gait wearable robots (AGWR) is expected to grow in the next few years to address the gait impairment of the aging population along with children and patients who suffer from neurological disorders. The gait of subjects wearing the assistive robots needs to be assessed objectively to improve the AGWR design, controllability and energy consumption. The objective assessment of the wearable robots will lead to the future self-adaptable and smart devices including exoskeletons, orthoses, and prosthetics that easily adapt to the user and environment requirement which is one of the main obstacle of routine use of such devices.

For that reason, in this chapter, the challenges and the major parameters in designing and evaluating the AGWR for indoor and the outdoor environment were presented. The assessment parameters were divided into two categories:

(1) Parameters leading to assessment of the interaction comfort level for the user while using the device,
(2) Parameters that assess the overall system functionality including the user and the device together.

Two major approaches for gait analysis employed in the design of assistive robots, based on vision systems and wearable sensors, were described. Exemplars of these approaches were accompanied by experiments for gait analysis with above-knee amputees and recognition of ADLs with healthy participants. Additionally, a list of essential design and performance parameters, such as comfort, safety, adaptability, compliance, and energy consumption, were presented for the development of efficient, adaptive and safe assistive robots.

From what is presented in this chapter, an important point to be considered is that there are currently limitations both in laboratory measurements and in the use of wearable sensors for measurements in outdoor environments. Successful designs of more intelligent wearable assistive robotic systems require accurate data achieved from real activities in outdoor environments. Some of these aspects are already under investigation across the world and promising results are being produced, however, many of these indices and methods of measurements still remain a major challenge for researcher and engineers.

References

1. Beauchet, O., Annweiler, C., Callisaya, M. L., De Cock, A.-M,, Helbostad, J. L., Kressig, R. W., et al. (2106). Poor gait performance and prediction of dementia: results from a meta-analysis. *Journal of the American Medical Directors Association, 17*, 482–490.
2. Cohen, J. A., Verghese, J., & Zwerling, J. L. (2016). Cognition and gait in older people. *Maturitas, 93,* 73–77.
3. Calderón-Garcidueñas, L., Mora-Tiscareño, A., Ontiveros, E., Gómez-Garza, G., Barragán-Mejía, G., Broadway, J., et al. (2008). Air pollution, cognitive deficits and brain abnormalities: A pilot study with children and dogs. *Brain and Cognition, 68,* 117–127.
4. Harada, M. (1995). Minamata disease: methylmercury poisoning in Japan caused by environmental pollution. *Critical Reviews in Toxicology, 25,* 1–24.
5. Burgar, C. G., Lum, P. S., Shor, P. C., & Van der Loos, H. M. (2000). Development of robots for rehabilitation therapy: The Palo Alto VA/Stanford experience. *Journal of Rehabilitation Research and Development, 37,* 663–674.
6. Huisinga, J. M., St George, R. J., Spain, R., Overs, S., & Horak, F. B. (2014). Postural Response Latencies Are Related to Balance Control During Standing and Walking in Patients With Multiple Sclerosis. *Archives of Physical Medicine and Rehabilitation, 95,* 1390–1397.
7. Horak, F. B. (2006). Postural orientation and equilibrium: what do we need to know about neural control of balance to prevent falls? Age and Aging, *35,* ii7–ii11.
8. Winter, D. A. (2009). *Biomechanics and Motor Control of Human Movement* (4th edn.). Wiley.
9. Zajac, F. E., Neptune, R. R., & Kautz, S. A. (2003). Biomechanics and muscle coordination of human walking: Part II: Lessons from dynamical simulations and clinical implications. *Gait & Posture, 17,* 1–17.
10. Herr, H. (2009). Exoskeletons and orthoses: Classification, design challenges and future directions,". *Journal of NeuroEngineering and Rehabilitation, 6,* 21–30.

11. Alireza, A., Martinez-Hernandez, U., Awad, M. I., Bradley, D. A., Dehghani-Sanij, A. A. (2018). Human-activity-centered measurement system: challenges from laboratory to the real environment in assistive gait wearable robotics. In *16th Mechatronics Forum International Conference* (pp. 38–43). University of Strathclyde Publishing.
12. Allard, P., Stokes, I. A., & Blanchi, J.-P. (1995). *Three-dimensional analysis of human movement*, Human Kinetics Publishers.
13. Sutherland, D. H. (2002). The evolution of clinical gait analysis: Part II Kinematics. *Gait & Posture, 16,* 159–179.
14. Patel, S., Park, H., Bonato, P., Chan, L., & Rodgers, M. (2012). A review of wearable sensors and systems with application in rehabilitation. *Journal of NeuroEngineering and Rehabilitation, 9,* 21–38.
15. Liu, T., Inoue, Y., & Shibata, K. (2009). Development of a wearable sensor system for quantitative gait analysis. *Measurement, 42,* 978–988.
16. Martinez-Hernandez, U., & Dehghani-Sanij, A. A. (2018). Adaptive Bayesian inference system for recognition of walking activities and prediction of gait events using wearable sensors. *Neural Networks, 102,* 107–119.
17. Maqbool, H. F., Muhammad, A. B. F., Awad, M. I., Abouhossein, A., Iqbal, N., & Dehghani-Sanij, A. A. (2017). A real-time gait event detection for lower limb prosthesis control and evaluation. *IEEE Transactions Neural Systems & Rehabilitation Engineering, 25,* 1500–1509.
18. Chen, B., Ma, H., Qin, L. Y., Gao, F., Chan, K. M., Law, S. W., et al. (2016). Recent developments and challenges of lower extremity exoskeletons. *Journal of Orthopaedic Translation, 5,* 26–37.
19. Esquenazi, A., Talaty, M., Packel, A., & Saulino, M. (2012). The ReWalk powered exoskeleton to restore ambulatory function to individuals with thoracic-level motor-complete spinal cord injury. *American Journal of Physical Medicine & Rehabilitation, 91,* 911–921.
20. Talaty, M., Esquenazi, A., & Briceno, J. E. (2013). Differentiating Ability in Users of the ReWalk (TM) Powered Exoskeleton An Analysis of Walking Kinematics. In *13th IEEE International Conference on Rehabilitation Robotics (ICORR)* (pp. 1–5). Seattle.
21. Farris, R. J., Quintero, H. A., Murray, S. A., Ha, K. H., Hartigan, C., & Goldfarb, M. (2014). A Preliminary Assessment of Legged Mobility Provided by a Lower Limb Exoskeleton for Persons With Paraplegia. *IEEE Transactions Neural Systems & Rehabilitation Engineering, 22,* 482–490.
22. Farris, R. J., Quintero, H. A., Withrow, T. J., & Goldfarb, M. (2009). *Design of a joint-coupled orthosis for FES-aided gait* (pp. 246–252). Kyoto: International Conference on Rehabilitation Robotics.
23. Murray, S. A., Ha, K. H., Hartigan, C., & Goldfarb, M. (2015). An Assistive Control Approach for a Lower-Limb Exoskeleton to Facilitate Recovery of Walking Following Stroke. *IEEE Transactions on Neural Systems & Rehabilitation Engineering, 23,* 441–449.
24. Ha, K. H., Murray, S. A., & Goldfarb, M. (2016). An Approach for the Cooperative Control of FES With a Powered Exoskeleton During Level Walking for Persons With Paraplegia. *IEEE Transactions on Neural Systems & Rehabilitation Engineering, 24,* 455–466.
25. Lajeunesse, V., Vincent, C., Routhier, F., Careau, E., & Michaud, F. (2016). Exoskeletons' design and usefulness evidence according to a systematic review of lower limb exoskeletons used for functional mobility by people with spinal cord injury. *Disability & Rehabilitation: Assistive Technology, 11,* 535–547.
26. REX Bionics. (2017). Robot for Rehabilitation: Exercising, Walking and Standing. @ www.rexbionics.com. Accessed May 20, 2019.
27. Sankai, Y. (2010). HAL: Hybrid assistive limb based on cybernics. *Robotics Research, 66,* 25–34.
28. Sanz-Merodio, D., Cestari, M., Arevalo, J. C. & Garcia, E. (2012). A lower-limb exoskeleton for gait assistance in quadriplegia. In *IEEE International Conference on Robotics and Biomimetics (ROBIO)* (pp. 122–127). Guangzhou.
29. Sancho-Perez, J., Perez, M., Garcia, E., Sanz-Merodio, D., Plaza, A., Cestari, M.: Mechanical Description of ATLAS 2020, A 10-DOF Paediatric Exoskeleton, *Advances in Cooperative Robotics*, 814–822 (2017).

30. Sanz-Merodio, D., Sancho, J., Perez, M., & Garcia, E. (2016). Control architecture of the ATLAS 2020 lower-limb active orthosis. In *Advances in Cooperative Robotics, Proceeding of 19th CLAWAR International Conference* (pp. 860–868). London,
31. Rossi, S., Patanè, F., Sette, F. D., Cappa, P. (2014). WAKE-up: A wearable ankle knee exoskeleton. In: *5th IEEE RAS/EMBS International Conference Biomedical Robotics & Biomechatronics* (pp. 504–507), Sao Paulo.
32. Patane, F., Rossi, S., Sette, F. D., Taborri, J., & Cappa, P. (2017). WAKE-up exoskeleton to assist children with Cerebral Palsy: design and preliminary evaluation in level walking. *IEEE Transactions on Neural Systems & Rehabilitation Engineering, 25,* 906–916.
33. Lusardi, M. M., Nielsen, C. C., Emery, M. J., Bowers, D. M., & Vaughan, V. G. (2007). *Orthotics and Prosthetics in Rehabilitation* (2nd ed). Elsevier.
34. Inman, V. T. (1966). Human locomotion. *Canadian Medical Association Journal, 94*(20), 1047–1054.
35. Franceschini, M., Massucci, M., Ferrari, L., Agosti, M., & Paroli, C. (2003). Effects of an ankle-foot orthosis on spatiotemporal parameters and energy cost of hemiparetic gait. *Clinical Rehabilitation, 17,* 368–372.
36. Traballesi, M., Porcacchia, P., Averna, T., & Brunelli, S. (2008). Energy cost of walking measurements in subjects with lower limb amputations: a comparison study between floor and treadmill test. *Gait & Posture, 27,* 70–75.
37. Vrieling, A. H., van Keeken, H. G., Schoppen, T., Hof, A. L., Otten, B., Halbertsma, J. P. K., et al. (2009). Gait adjustments in obstacle crossing, gait initiation and gait termination after a recent lower limb amputation. *Clinical Rehabilitation, 23,* 659–671.
38. Sapp, L., & Little, C. (1995). Functional outcomes in a lower limb amputee population. *Prosthetics and Orthotics International, 19,* 92–96.
39. Bruijn, S., Meijer, O., Beek, P., & Van Dieën, J. (2013). Assessing the stability of human locomotion: a review of current measures. *Journal of the Royal Society Interface, 10*(83), 1–20.
40. Mahmood, I., Martinez, Hernandez, U., & Dehghani-Sanij, A. A. (2018). Gait dynamic stability analysis for simulated Ankle-foot impairments and Bipedal robotics application. In *16th Mechatronics Forum International Conference* (pp. 58–64). Strathclyde.
41. Sadeghi, H., Allard, P., Prince, F., & Labelle, H. (2000). Symmetry and limb dominance in able-bodied gait: A review. *Gait & Posture, 12*(1), 34–45.
42. Andres, R., & Stimmel, S. (1990). Prosthetic alignment effects on gait symmetry: A case study. *Clinical Biomechanics, 5*(2), 88–96.
43. Cabral, S., Resende, R. A., Clansey, A. C., Deluzio, K. J., Selbie, W. S., & Veloso, A. P. (2016). A global gait asymmetry index. *Journal of applied biomechanics, 32*(2), 171–177.
44. Martinez-Hernandez, U., Mahmood, I., & Dehghani-Sanij, A. A. (2017). Simultaneous bayesian recognition of locomotion and gait phases with wearable sensors. *IEEE Sensors Journal, 18*(3), 1282–1290.
45. Varol, H. A., Sup, F., & Goldfarb, M. (2010). Multiclass real-time intent recognition of a powered lower limb prosthesis. *IEEE Trans. Biomedical Engineering, 57*(3), 542–551.
46. Jimenez-Fabian, R., & Verlinden, O. (2012). Review of control algorithms for robotic ankle systems in lower-limb orthoses, prostheses, and exoskeletons. *Medical Engineering & Physics, 34*(4), 397–408.
47. Martinez-Hernandez, U., & Dehghani-Sanij, A. A. (2018). Probabilistic identification of sit-to-stand and stand-to-sit with a wearable sensor. *Pattern Recognition Letters, 118,* 32–41.

A High Fidelity Driving Simulation Platform for the Development and Validation of Advanced Driver Assistance Systems

Marco Grottoli, Anne van der Heide and Yves Lemmens

Abstract New vehicle designs with advanced driver assistance systems need to be validated with respect to human perceptions of comfort and risk. Therefore, human-in-the-loop simulations are used to evaluate a wide range of scenarios in driving simulators. In order to improve human-in-the-loop simulation, the chapter begins by reporting solver advancements that enable the real-time simulation of complex mechatronic systems using high fidelity multibody and multi-physics simulation models. A driving simulator setup is then presented that makes use of the high-fidelity vehicle models and can simulate vehicles with advanced driver assistance systems. The essential components of the simulator are outlined and initial results of a comparison study between high fidelity model and equivalent low fidelity models. Finally, two test cases are described that use respectively an adaptive cruise control function and an autonomous intersection crossing function.

List of Abbreviations

ABS	Anti-Lock Braking System
ACC	Adaptive Cruise Control
ADAS	Advanced Driver-Assistance Systems
ADF	Automated Driving Functionality
AIC	Autonomous Intersection Crossing
CFD	Computational Fluid Dynamics
DoF	Degree(s) of Freedom
DTM	Double Track Model
EPS	Electric Power Steering
ESP	Electronic Stability Program
HiL	Hardware in the Loop
HuiL	Human in the Loop
ICE	Internal Combustion Engine

M. Grottoli · A. van der Heide · Y. Lemmens (✉)
Siemens PLM Software, Louvain, Belgium
e-mail: yves.lemmens@siemens.com

X.-T. Yan et al. (eds.), *Reinventing Mechatronics*,
https://doi.org/10.1007/978-3-030-29131-0_7

MBS Multibody Simulation
MCA Motion Cueing Algorithm
OEM Original Equipment Manufacturer
SiL Software in the Loop
STM Single Track Model

Introduction

During a design cycle, many simulation models are built to analyse the behaviour and performance of a mechatronic system. Often, different simulation models of the same system are built by different engineering departments. A unified modelling process is therefore often sought that aims to reuse similar simulation models in order to speed up the design process and make more efficiently use of resources. It is a practice that is already actively used by some automotive OEMs [1]. However, due to conflicting requirements, the reuse of models is not always possible. For example, vehicle models for driving simulators need to run in real-time to enable a realistic interaction with a driver. Therefore, these models traditionally consist of a series of relatively simple analytical formulae or look-up tables derived from more detailed physical simulation models from other departments. Employing these reduced order models is a cumbersome process since new models need to be created and correlated every time a design change occurs.

On the other hand, digital or virtual prototyping using physics-based models have become a standard part of the detailed design process of mechatronic systems. For example, multibody simulation models (MBS) of a car suspension are used to predict the ride and handling qualities of the vehicle. These MBS models are usually coupled with models of those vehicle systems that affect the driving dynamics such as the powertrain or Electronic Stability Programmes. These high-fidelity models are often correlated with test data to increase accuracy.

The chapter first reports on recent technological advances made by Siemens PLM Software® in its simulation tools for multibody analyses and multi-physics simulation in order to run detailed models in real-time. These advances include the development of a new deterministic solver that can handle the same models as the traditional off-line solver. This not only enables the reuse of models for real-time simulation, but the real-time simulation results will be more accurate when compared to the reduced order models. Moreover, design parameters can be changed directly in the real-time simulation models while all internal variables are available during and after the real-time simulation. These advances are summarised below.

The chapter then describes a test setup of a driving simulator using high fidelity vehicle models that can be used to validate advanced driver-assistance systems (ADAS) with respect to human perceptions of comfort and risk. Following sections then describe the system architecture and the essential components of the simulation platform. A comparison is then made between the multibody vehicle model used in the simulator and the lower fidelity vehicle dynamic models. The simulation platform is then discussed in relation to two automated driving functionalities, namely adaptive cruise control and autonomous intersection crossing. Conclusions and future work are presented in the final section of the paper.

Real-Time Simulation

In real-time simulation it is of vital importance that the turnaround time of a time step, i.e. the time it takes to compute the system states at the next discretized point in time, is a priori known to be less than the simulated time step. The multibody analyses and multi-physics simulation software considered uses implicit time integration schemes which make sure that the accelerations satisfy the model equations at the next discretized time point. However, they require an unknown number of iterations to reach a converged solution. Therefore, the number of iterations has been limited and the order of the time integration scheme fixed to achieve deterministic behaviour of the solver as proposed by Garcia de Jalon and Bayo [2].

Additionally, every iteration step in every time step of an implicit time integration scheme involves the evaluation of the system Jacobian, and the factorization of this matrix. Classically, the Jacobian is obtained by finite difference methods, with considerable computational time costs. To overcome this, analytical expressions for the Jacobian have been implemented in multibody simulation software Simcenter™ 3D Motion, reducing the cost of Jacobian evaluation dramatically.

Another source of variable computation cost, is the drift or violation of the constraint equations that can occur during the solution process. As in real-time analysis this constraint equation correction must be done efficiently, Baumgarte's stabilization method has been included in the multibody solver to correct the drift [3]. The advantage of Baumgarte's method is that it does not add much computational load since the sparsity pattern of the Jacobian does not change.

Finally, the duration of simulations can be reduced through the parallelisation of parts of the simulation. This is already common practice for many analysis tools, as for instance Computational Fluid Dynamics (CFD) simulation tools. One option is a low-level approach where the algorithm is divided into multiple parts. However, this is a non-trivial task in multibody dynamics due to the coupling between the degrees-of-freedom due to constraints and force elements such as bushings, springs and dampers. Therefore, a more high-level approach has been taken that subdivides the model such that co-simulation on separate cores can be used. In this approach, the model is split into two parts and each part is then solved on a different processor core. In case of system simulation, the co-simulation interface can be used to exchange parameter states at the interface between the two mechanisms during every time step. In case of multibody simulation, a dedicated element has been developed that is based on a stiff bushing element to connect 2 sub-models together. Because an implicit time integration scheme can handle the resulting higher numerical stiffness, a stiff bushing is not penalizing for the computational effort. [4].

Components of the Human-in-the-Loop Simulation Platform

In order to investigate, demonstrate and research multibody vehicle models and sub-system models for real-time Software in the Loop/Hardware in the Loop/Human in the Loop (SiL/HiL/HuiL) capabilities, a test setup for a driving simulator has been built. The complete simulation platform is comprised of several key components. These include the real-time models of the vehicle and its critical sub-systems for simulating vehicle behaviour, a virtual environment that can interact with vehicle model, control algorithms, sensor models, and communication models, and a real-time co-simulation platform for deterministic computation. Moreover, appropriate tactile, motion, acoustic and visual feedback is crucial for immersive driver-in-the-loop capabilities. Last, but not least, appropriate hardware is necessary both for the manual driving mode and the autonomous driving mode. The resulting system architecture for the simulation platform is shown in Fig. 1. In what follows, the various components are discussed in detail.

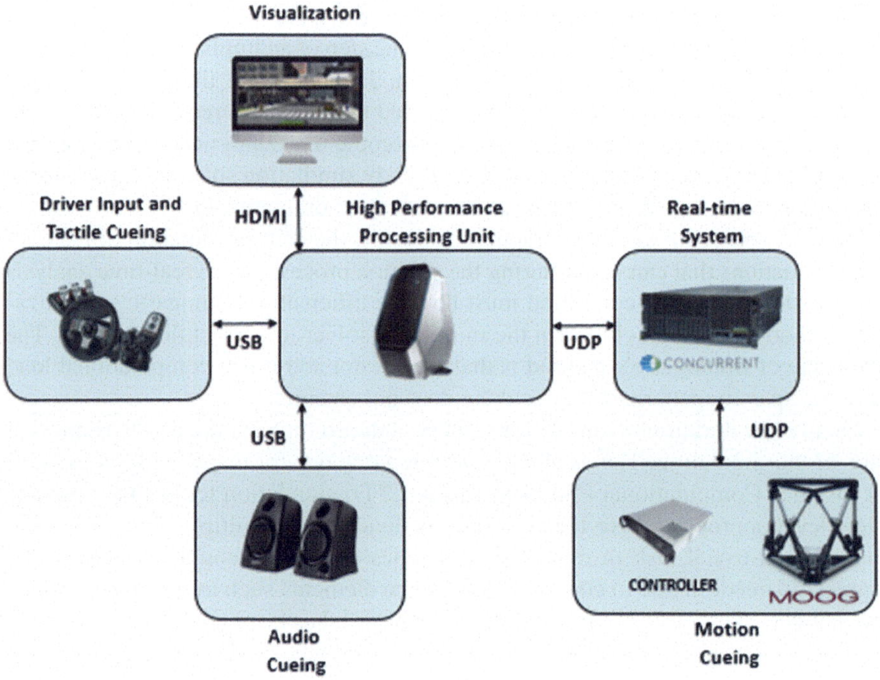

Fig. 1 System architecture for the driving simulation platform

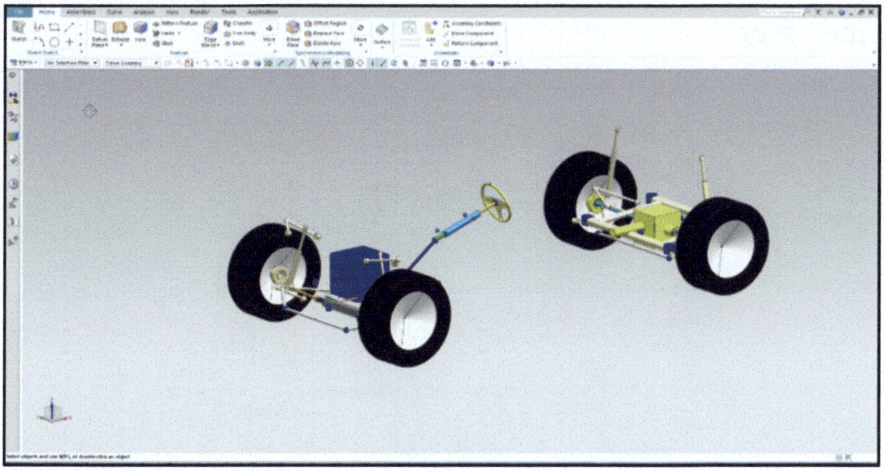

Fig. 2 Multibody model of the rear-wheel driven vehicle

Real-Time Vehicle Models

Multibody Model of a Vehicle

To model the realistic dynamic behaviour of the vehicle, such as ride and handling characteristics, a multibody modelling approach is used. An extensive survey of the (flexible-) multibody modelling approach is presented by Shabana [5, 6]. This implies that the complete vehicle behaviour can be modelled by interconnected rigid or flexible bodies, each of which may undergo large translational and rotational displacements. However, deriving the equation of motion manually for three-dimensional motion involving multiple bodies is tedious and cumbersome. For this reason, a real-time multibody model in Simcenter™ 3D Motion was developed. This model has 161 degrees-of-freedom (Fig. 2) which comprises of the body of the vehicle, engine body, driveline, front and rear suspensions and stabilize bars, steering mechanism and wheels. The input for this model are the torques at the four wheels (traction and braking torque) and the steering angle at the steering wheel.

Powertrain with Internal Combustion Engine

Next, a model for a powertrain with an internal combustion engine, shown in Fig. 3, is developed. The complete model of the powertrain comprises an internal combustion engine (ICE) with an electronic control unit for regulation, a manual gearbox including clutch, a differential rear-axle with engine feedback. The model of ICE is capable of computing the engine torque, fuel consumption, thermal loses, exhaust gas flow rate and emissions such as CO, NO_x, HC and particulates. The model of

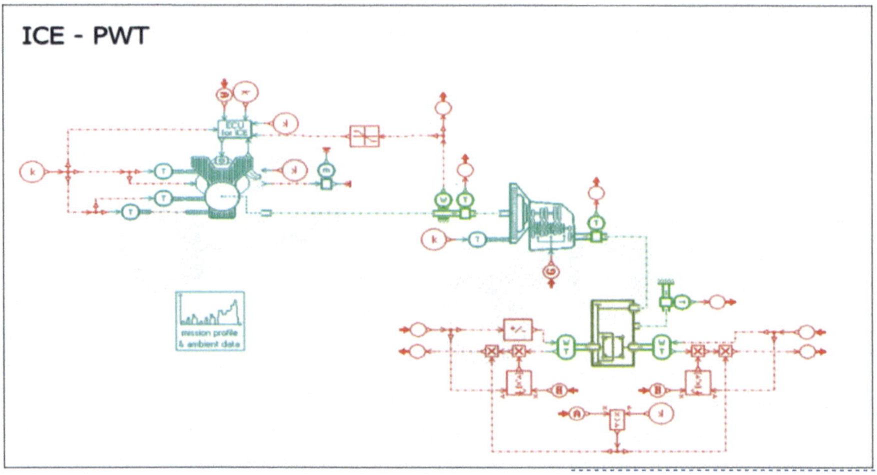

Fig. 3 Model of the powertrain with ICE

ICE can be appended to any drivetrain. In this case, it will operate with a manual gear box which will ultimately transmit power to the differential rear-axle.

Anti-lock Braking System (ABS)/Electronic Stability Program (ESP)

The safety of drivers, passengers, other road users and infrastructure are always a priority. To this end, modern vehicles are equipped with state-of-the-art automobile safety systems such as antilock braking systems (ABS), antiskid regulation (ASR) and an electronic stability program (ESP). In this work, a lumped parameter approach is used to express these safety systems. The ESP improves vehicle handling and control in critical situations. This is achieved by applying appropriate braking forces to each wheel of the vehicle. Whereas, as their names suggest, ABS and ASR allow the wheels to preserve tractive contact with the road while braking by preventing the wheels from locking and slipping. In conjunction with the ABS and ASR, ESP (as shown in Fig. 4) improves vehicle stability and control and consequently make the vehicle safer to drive.

Electric Power Steering (EPS)

In order to reduce driver fatigue, modern vehicles are equipped with power steering systems using either a hydraulic, an electro-hydraulic or an electrical system. As electrical systems require fewer auxiliary components (belt driven mechanism, hydraulic pump etc.) and are safer and cleaner when compared to their hydraulic counterpart, an electrical power assisted steering is modelled as shown in Fig. 5.

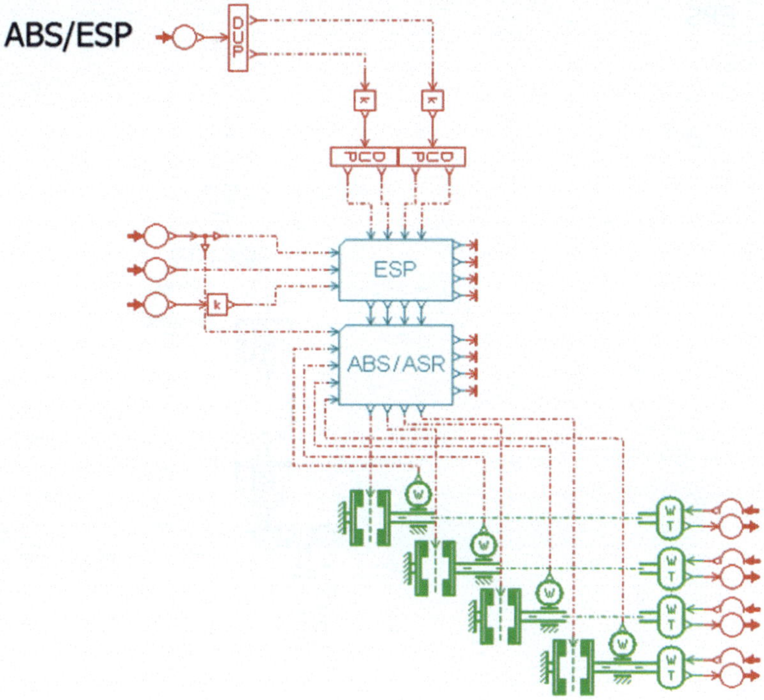

Fig. 4 Model of ABS and ESP

Virtual Environment, Sensor and Communication Models

To simulate or test an Advanced Driver Assistance Systems (ADAS) functionality with driver-in-the-loop capabilities, the modelling of the virtual world is indispensable. To this end, a physics based simulation solution Simcenter® Prescan was used to design scenarios (e.g. ground plan, road conditions and infrastructure etc.), model the environment (e.g. pedestrians, other vehicles/obstacles/objects and weather modules), sensors (such as radar, laser/LIDAR, camera and GPS) and communication technologies (as for instance vehicle-to-vehicle and vehicle-to-infrastructure) for evaluating ADAS. Here, different scenarios, sensor models and communication methods are used for the purpose of demonstration.

Automated Driving Functionality

In order to perform the closed-loop simulation, besides the vehicle model and other simulation components, an appropriate driving functionality is also required. In this

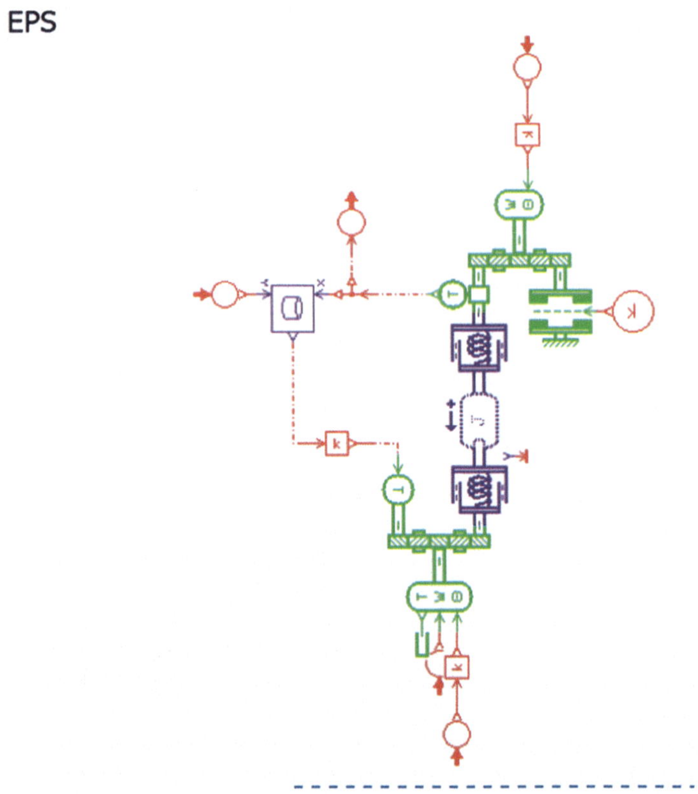

Fig. 5 Model of electric power steering

paper, two automated driving functionalities (ADFs) are investigated and implemented for the purpose of demonstration. Although the existing state of these controllers is at an infancy level, in what follows, a brief outline is given.

Adaptive Cruise Control (ACC)

The first ADF investigated and used on the simulator is adaptive cruise control (ACC). This is an automatic control that helps the driver to maintain a safe distance with the respect to a leading car. The aim of the control is to avoid collision by automatically releasing the throttle and/or actuating the brake pedal of the ego vehicle under test with autonomous functionality [7]. The actions taken by the ACC control algorithm are based upon the information given by two scanning sensors: long range radar (LRR) 150 m range with narrow beam of 9° and short-range radar (SRR) with 30 m detecting range and a wide beam of 80°. The sensors continuously evaluate the presence of a lead car. In the case a slower car gets inside the sensor measuring range, the control procedure computes the relative distance and speed between the two vehicles. The brake system is triggered once the relative distance and relative speed between the ego and lead car goes over a pre-defined value.

In case the relative speed is around zero, the control will release the throttle pedal. In the case when safe distance is violated and the cars relative speed is above a predefined threshold, the control procedure acts on the ego brake by modulating its intensity in order to maintain the safe distance. The extensive details of the complete ACC algorithm used for the current work is beyond the scope of the paper and hence omitted. However, different ACC algorithms can be found in the literature, as for instance those by Stanger and del Re [8] and Naranjo et al. [9].

Autonomous Intersection Crossing (AIC)

The second ADF implemented on the simulator is a controller for autonomous intersection crossing (AIC) that can only consider one other vehicle. This functionality comprises a pure pursuit controller and collision avoidance. The pure pursuit algorithm is a classical path tracking algorithm and enforces the ego vehicle to move to a look-ahead point ahead of it from its current position on a pre-defined trajectory. The algorithm then changes the look-ahead point on the trajectory depending on the existing position of the vehicle [10]. For the collision avoidance at the intersection, an algorithm presented by del Vecchio et al. [11] and Hafner et al. [12] is adopted and implemented. In this algorithm, it is assumed that both vehicles are following a predefined trajectory, which makes the problem solvable by only changing longitudinal velocity. If a collision is predicted at the current speed of the ego vehicle, the algorithm forces the ego vehicle to either brake or accelerate depending on the states of both the vehicles to avoid the collision. This is achieved by defining a certain *bad set* which is the set of all velocities that will result in a collision. The process involves the generation of a feedback map that prevents the flow of vehicles from entering the *bad set* and consequently avoiding a collision. In this way, one vehicle will try to exit the intersection as fast as possible and the other will try to slow down as much as possible to avoid a collision.

Motion Cueing

Inertial or motion cues play a significant role for the achievement of an immersive feeling in motion driving simulators (Fig. 6). Therefore, the driving simulator is mounted on a hexapod with 6 Degrees of Freedom (DoF). Ideally, the inertial motion provided by the motion system should be the same as that experienced by a real driver. However, due the physical limitations of the adopted motion system, the resulting inertial motion differs from the ideal case. The Motion Cueing Algorithm (MCA) is a control logic system which generally takes as input the linear accelerations and the angular velocities of the simulated vehicle and computes the reference signal (in terms of positions, velocities and accelerations) for the motion system. Different

Fig. 6 Driving simulator installation with 6 DoF hexapod

MCAs exists for driving simulators [13–15]. The one used here is known as Classical MCA [16] and is one of the most generally adopted algorithms. Additionally, MCA based on model predictive controller strategies is also being investigated.

Comparison of Vehicle Dynamics Models

In order to evaluate the effect of the fidelity of the multibody vehicle model that is used in the simulator, its dynamics behaviour is compared to the traditional lower fidelity models with off-line simulations. In this case, single track and double track models available in the multi-physics simulation software Simcenter Amesim™ are used for comparison.

Simplified Models

The first dynamics model considered for comparison is the single track model (STM) of Fig. 7. This consists of a rigid body which moves on the horizontal plane and is comprised of two tyres, one in the front and one in the rear. The tyre model computes the forces exchanged with the road and the forces applied to the chassis. The STM cannot provide an estimation of pitch or roll of the vehicle, but provides an estimation of the vehicle's yaw and therefore it is widely used in vehicle dynamics simulations. The inputs for this model are the steering angle at the front wheel and vehicle velocity.

The second lateral dynamics model considered in this study is a double track model (DTM) of Fig. 8. The DTM has a similar setup to the STM, but has two extra wheels. The positions of the four wheels resemble those of a real vehicle, and therefore it is possible to capture the load transfer to the tyres during lateral manoeuvres, and consequently the chassis roll angle. The tyre model used has a corning stiffness which depends on the vertical load but is limited to a maximum stiffness. Also, the steering mechanism is considered to be linear, and the steering angle is directly input at the wheel. The geometry, mass and inertia of the vehicle are set to the same values of the high fidelity model. The final parameter to set is the roll stiffness. This value is found by making use of the stiffness of the suspensions elements and the location of the centre of roll. It is assumed that the suspension elements are placed in a vertical orientation and the centre of roll is at the same location as the centre of mass. The inputs to the vehicle model are the steering wheel angle and the longitudinal velocity. The results obtained from the model are compared with the results of the STM and with the multibody model for the lateral dynamics scenarios.

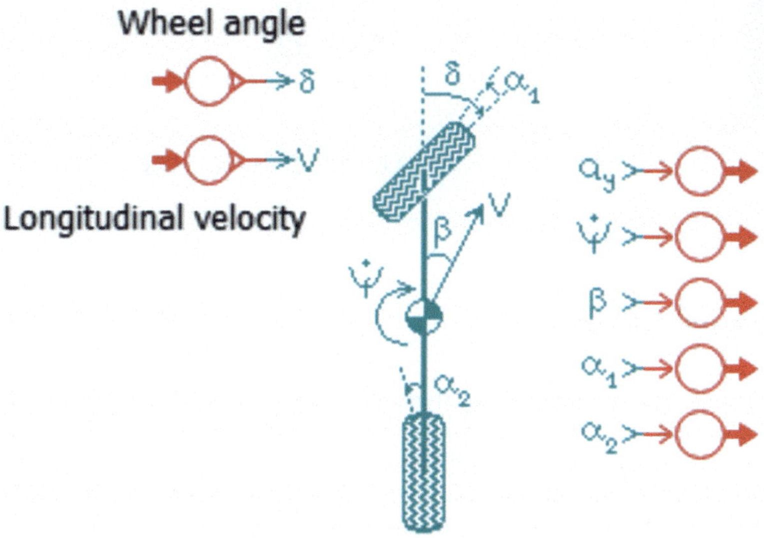

Fig. 7 Single track model

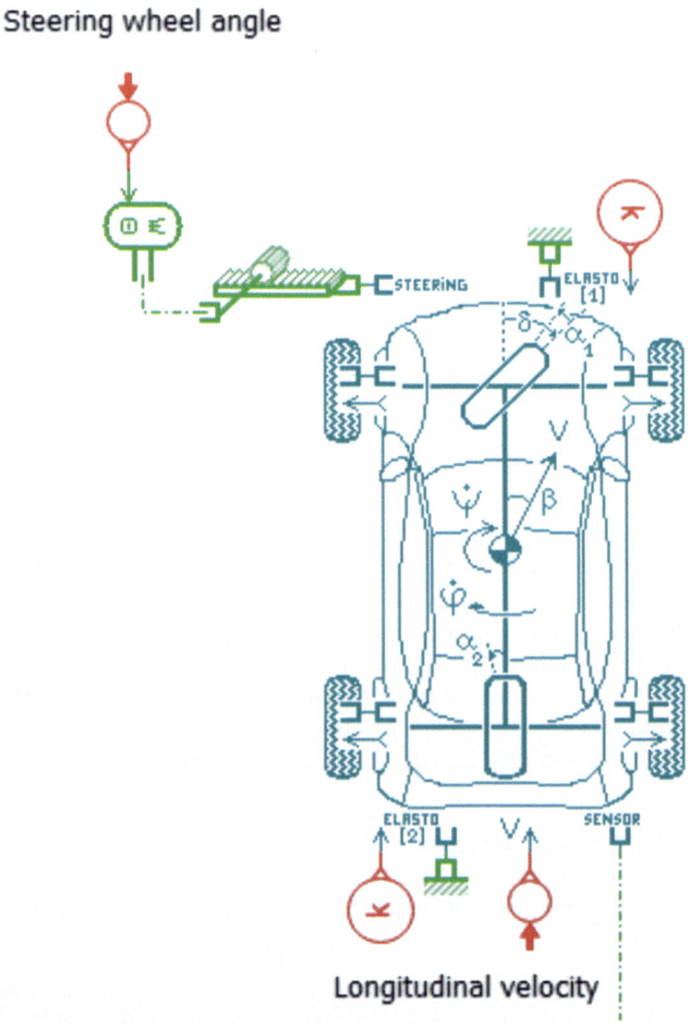

Fig. 8 Double track model

Comparison Studies

The first results are for a longitudinal dynamics scenario, in which braking is activated by the ACC while the vehicle is moving at the constant speed of 50 km/h and performs the braking manoeuvre until standstill. The resulting velocity profiles and pitch angles obtained with the multibody model and the STM are shown in Fig. 9 when Fig. 9a shows the velocity profiles and Fig. 9b the pitching angle.

From Fig. 9a it can be seen that both velocity profiles are similar. However, the pitching angle obtained by the simplified models differs from that for the multibody

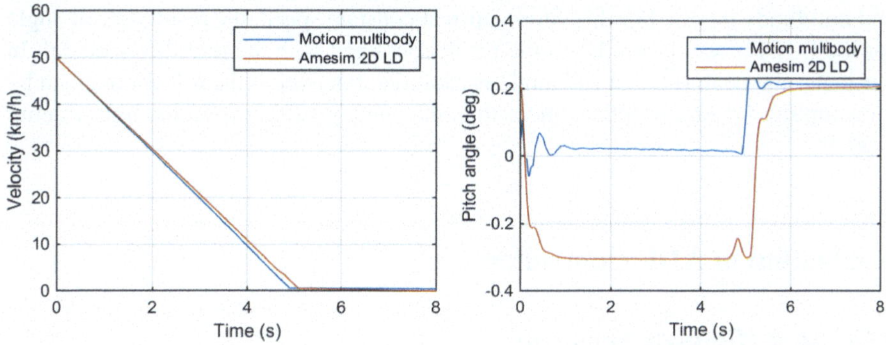

(a) Velocity profiles (b) Pitching angle

Fig. 9 Velocity profile and pitch angle for the longitudinal dynamics scenario. **a** Velocity profiles, **b** Pitching angle

model. This is due to the simplified suspension system used with the longitudinal dynamics model with the geometry of the actual suspension system being far more complex than the vertical springs are assumed in the simplified model. However, the difference is limited in this case due to low values of deceleration. For an extreme manoeuvre like emergency braking with more non-linear behaviour, it is anticipated that the differences will be greater. The difference in total braking distance between both models is 1.5 m.

The second scenario considered is a turn performed at a constant speed of 20 km/h. For this scenario, the resulting trajectories are compared for three models: the multibody model, the STM and the DTM. As can be seen in Fig. 10a, the resulting trajectories for the 3 models are equal.

The final comparison in this manoeuvre is the roll angle of the vehicles. Because the STM does not include roll motion, Fig. 10b only shows the results of the DTM

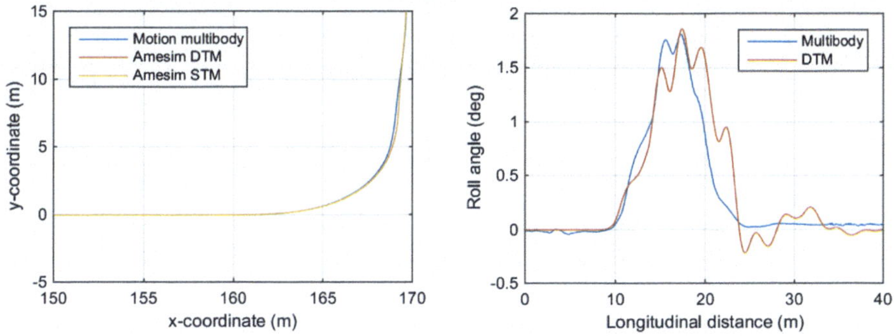

(a) Trajectory (b) Roll angle

Fig. 10 Trajectory and roll angle for the lateral dynamics manoeuvre. **a** Trajectory, **b** Roll angle

and multibody model. For this simple turn at constant speed, the resulting roll angle for the DTM model is similar to the results obtained with the multibody model. In particular the peak value of roll angle is captured, but overall the roll angle given by the simplified model oscillates more than the angle resulting from the high-fidelity model.

Evaluation of ADF Controllers

ACC on a Highway Scenario

For this particular test case, a section of E40 highway scenario near Bertem in Belgium is virtualised using real world map data. The virtual scenario, sensor models and traffic simulation are all developed using Simcenter® Prescan. A dedicated button on the steering wheel is used for enabling/disabling the ACC functionality. In this scenario, the ego vehicle is cruising at approximately 100 km/h and approaching a leading vehicle. To avoid the collision, the ego vehicle slows down to the same speed as the leading vehicle as is shown in Fig. 11. After 41 s, the ego vehicle decides to overtake the leading vehicle. From this moment, the ACC increases the velocity of the ego to the previously defined setpoint.

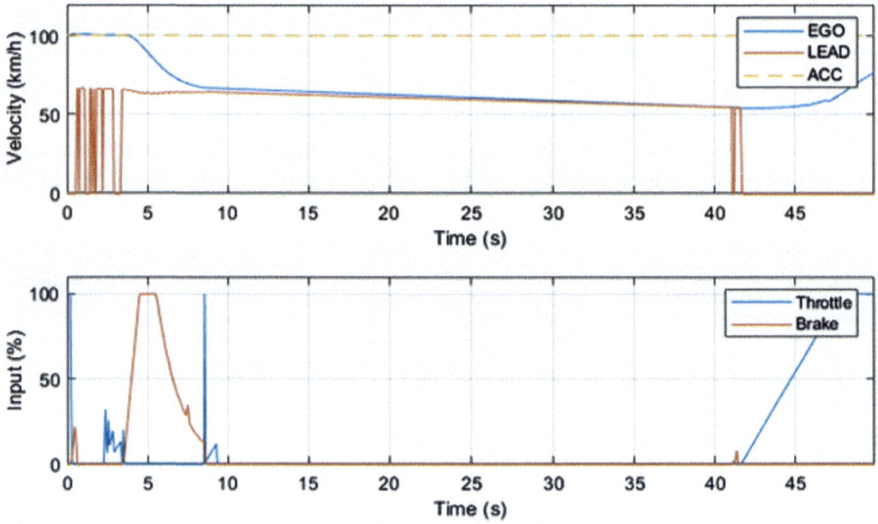

Fig. 11 Demonstration of the ACC on a highway scenario

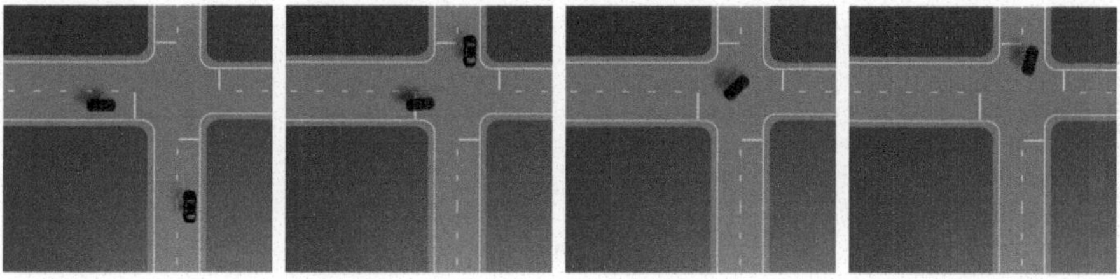

Fig. 12 Visualization of the AIC

Autonomous Intersection Crossing

The autonomous intersection crossing controller is demonstrated in the second scenario. The developed control algorithm is validated by having the ego vehicle approaching a cross-road together with one other road user. At a certain moment, their trajectories are overlapping and a collision might occur. Both vehicles are making a left turn and there is no communication between the road users. In this scenario, the ego vehicle arrives first at the intersection. It decides to slow down and give the priority to the other vehicle, because there is a chance of collision. After the road user leaves the intersection, the ego vehicle accelerates and makes a left turn. The scenario is visualized in Fig. 12 and the velocities and control inputs are shown in Fig. 13. The ego vehicle is shown by the red vehicle model and the other road user by the black.

Conclusions & Outlook

Mechatronic systems are getting more complex and more autonomous, clearly demonstrated by the trends in the automotive industry. As a result, predicting the accurate behaviour of autonomous vehicles in all possible circumstance is becoming very challenging. The failure to anticipate limitation in the perceptions of obstacles and response strategies have already resulted to grave accidents [17]. Moreover, the evaluation of automated driving functions in terms of the perceived comfort and risk by the occupants is becoming an important aspect in their development. Therefore realistic driving simulators that make use of high-fidelity simulations models are becoming important.

On the other hand, the reuse of models is also becoming important to speed up the development process. Therefore, this chapter reports on efforts to adapt multibody and multi-physics solvers to enable the reuse of offline vehicle dynamics models as real-time models for a driving simulator that can be used for the subjective evaluation of the perceived comfort and risk of advance driving assistant systems. The results of an initial comparison of high fidelity multibody models with lower fidelity system models show that the simplified models perform quite well for the reproduction of

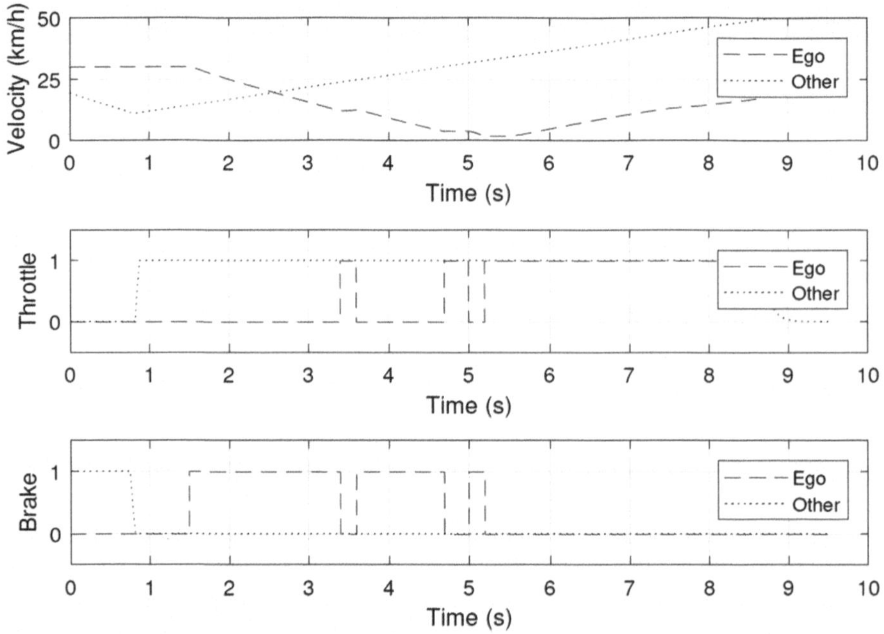

Fig. 13 Velocity and control inputs of the two vehicles approaching the intersection

vehicle trajectories and velocity profiles. The estimation of the roll and pitch angles, although less accurate, are still comparable with the results of the multibody model. From the analysis of the results it can be concluded that, as long as the manoeuvres to simulate are within the boundary conditions of the simplified models, the results are comparable. Further investigation will be done for more advanced manoeuvres, in which the non-linear dynamics of the vehicle will be more relevant and the differences between high fidelity and simplified models will be more evident.

The final test cases demonstrate that the high fidelity multibody and multi-physics models can be used for human-in-the-loop simulations and moreover that they verify the functional operation of the driving simulator setup. Further demonstration and validation activities of the simulator platform are anticipated. Firstly, the validation of the motion of the simulator platform with respect to a real vehicle is planned. Further, the development of methods for the subjective evaluation of perceived comfort and risk of ADAS in vehicles will be investigated. Finally, methods for the subjective evaluation of ride and handling qualities of manual driving vehicles are planned.

Acknowledgements The authors would like to acknowledge the ENABLE-S3 project that has received funding from the ECSEL Joint Undertaking under grant agreement No. 692455. This joint undertaking receives support from the European Union's HORIZON 2020 research and innovation programme and from the governments of Spain, Portugal, Poland, Ireland, Belgium, France, Netherlands, United Kingdom, Slovakia, Norway.

References

1. Prescott, W., Heirman, G., Furman, M., De Cuyper, J., Lippeck, A., & Brauner, H. (2012). Using high-fidelity multibody vehicle models in real-time simulations. In *SAE 2012 World Congress & Exhibition*, Detroit, Technical Paper No. 2012-01-0927.
2. Garcia de Jalon, J., & Bayo, E. (1994). *Kinematic and dynamic simulation of multibody systems: The real-time challenge*. New York: Springer.
3. Arnold, M., Burgermeister, B., & Eichberger, A. (2007). Linearly implicit time integration methods in real-time applications: DAEs and stiff ODEs. *Multibody System Dynamics, 17*(2–3), 99–117.
4. Prescott, W. (2011). Parallel processing of multibody systems for real-time analysis. In *Proceedings of the 2011 ECCOMAS Thematic Conference on Multibody Dynamics*, Brussels.
5. Shabana, A. A. (1997). Flexible multibody dynamics: Review of past and recent developments. *Multibody System Dynamics, 1,* 189–222.
6. Shabana, A. A. (2013). *Dynamics of multibody systems*. Cambridge: Cambridge University Press.
7. Hu, Y., Zhan, W., & Tomizuka, M. (2018). Probabilistic prediction of vehicle semantic intention and motion. In *IEEE Intelligent Vehicles Symposium* (pp. 307–313).
8. Stanger, T., & del Re, L. (2013). A model predictive cooperative adaptive cruise control approach. In *Proceedings of the American Control Conference* (pp. 1374–1379).
9. Naranjo, J. E., González, C., García, R., & de Pedro, T. (2006). ACC + Stop&Go maneuvers with throttle and brake fuzzy control. *IEEE Transactions on Intelligent Transportation Systems, 7*(2), 213–225.
10. Coulter, R. (1990). *Implementation of the pure pursuit path tracking algorithm*. Carnegie Mellon University Robotics Inst., Report CMU-RI-TR-92-01.
11. Del Vecchio, D., Malisoff, M., & Verma, R. (2009). A separation principle for hybrid automata on a partial order. In *Proceedings of the American Control Conference* (pp. 3638–3643).
12. Hafner, M. R., Cunningham, D., Caminiti, L., & Del Vecchio, D. (2013). Cooperative collision avoidance at intersections: Algorithms and experiments. *IEEE Transactions on Intelligent Transportation Systems, 14*(3), 1162–1175.
13. Schmidt, S. F., & Conrad, B. (1970). *Motion drive signals for piloted flight simulators*. NASA Technical Report @ ntrs.nasa.gov/archive/nasa/casi.ntrs.nasa.gov/19700017803.pdf. Accessed May 20, 2019.
14. Reid, L. D., & Nahon, M. A. (1985). *Flight simulator motion-base drive algorithms: Part 1—Developing and testing the equations*. Technical Report, University of Toronto Institute for Aerospace Studies.
15. Dagdelen, M., Reymond, G., Kemeny, A., Bordier, M., & Maïzi, N. (2004). MPC-based motion cueing algorithm: Development and application to the ULTIMATE driving simulator. In *Driving Simulation Conference (DSC)*, Paris (pp. 221–233).
16. Fang, Z., & Kemeny, A. (2012). Motion cueing algorithms for a real-time automobile driving simulator. In *Driving Simulation Conference (DSC)*, Paris (pp. 1–12).
17. en.wikipedia.org/wiki/List_of_self-driving_car_fatalities. Accessed May 20, 2019.

Advancing Assembly Through Human-Robot Collaboration: Framework and Implementation

Abdullah Mohammed and Lihui Wang

Abstract The chapter presents a framework for establishing human-robot collaborative assembly in industrial environments. To achieve this, the chapter first reviews the subject state of the art and then addresses the challenges facing researchers. The chapter provides two examples of human-robot collaboration. The first is a scenario where a human is remotely connected to an industrial robot, and the second is where a human collaborates locally with a robot on a shop floor. The chapter focuses on the human-robot collaborative assembly of mechanical components, both on-site and remotely. It also addresses sustainability issues from the societal perspective. The main research objective is to develop safe and operator-friendly solutions for human-robot collaborative assembly within a dynamic factory environment. The presented framework is evaluated using defined scenarios of distant and local assembly operations when the experimental results show that the approach is capable of effectively performing human-robot collaborative assembly tasks.

Introduction

Many international organizations and countries recognize the importance of sustainable development and hence of the need for defined strategies and policies which contribute to the directions of sustainability. Therefore, it is critical for companies situated to work toward cleaner production, and hence to be ready when regulations are activated.

The development consists of several directions, of which sustainable manufacturing is a major element. One of the more precise definitions of sustainable manufacturing is that from the US Department of Commerce [1] which reads:

> The creation of manufactured products that use processes that minimize negative environmental impacts, conserve energy and natural resources, are safe for employees, communities, and consumers and are economically sound.

A. Mohammed (✉) · L. Wang
KTH Royal Institute of Technology, Stockholm, Sweden
e-mail: agmo@kth.se

X.-T. Yan et al. (eds.), *Reinventing Mechatronics*,
https://doi.org/10.1007/978-3-030-29131-0_8

This means that the different fields of manufacturing need to be improved and aligned with the objectives of sustainability. A clear explanation of these aspects is reported by the OECD [2] which considers the three directions of sustainability of economy, society and environment.

Based on the above description of the sustainable manufacturing, the following need to be addressed to achieve the objectives of this chapter:

1. *Innovation*—The following innovative solutions are reported:
   - Introducing a 3D model-driven robotic system that allows a remote operator to control a robot to perform a series of assembly steps. Using a single camera mounted on the robot end-effector, the system is able to identify unknown objects within the robot's workspace and automatically incorporate them into a virtual 3D environment to generate tools for performing remote assembly operations. The system has been deployed with a physical robot and evaluated with basic assembly tasks [3].
   - Introducing a system to effectively detect and avoid any potential collision between humans and robots in. order to provide a safe and collaborative robotic environment. The system has been deployed and examined in a physical setup [4].
2. *Working conditions*—Improving the working conditions of shop floor operators was one of the main intentions of the work reported in the chapter. Therefore, safety was one of the major topics to be addressed by the research. As shown in Fig. 1, the highest priority is given to the human's safety. This is realized by using a set of depth sensors in a robotic cell to monitor the human operators and to control robots to avoid any potential collision.

Based on the characterization scheme proposed by Parasuraman et al. [6], the human-robot collaboration is focused on Level 8 of automation and interaction between the human and the robot as expressed by Fig. 2. At this level, a user interface is used to inform the human of the current status of the system. The system then reacts when a potential danger is detected in. order to protect the human. This means that the system will provide the collaboration automatically by showing the decision-making required to avoid collisions and alerting the operator if necessary. The operator can monitor the status of the system at runtime.

To establish collaboration between a human and a robot, a framework is introduced to support two types of human operators. The first is a remote operator; and the second a local operator on the shop floor. The following sections explain the related work strategies for these two individuals.

Human-Robot Remote Collaboration (HRRC)

In recent years, researchers have developed various tools to program, monitor and control industrial robots. The aim is through simulation to reduce potential robot

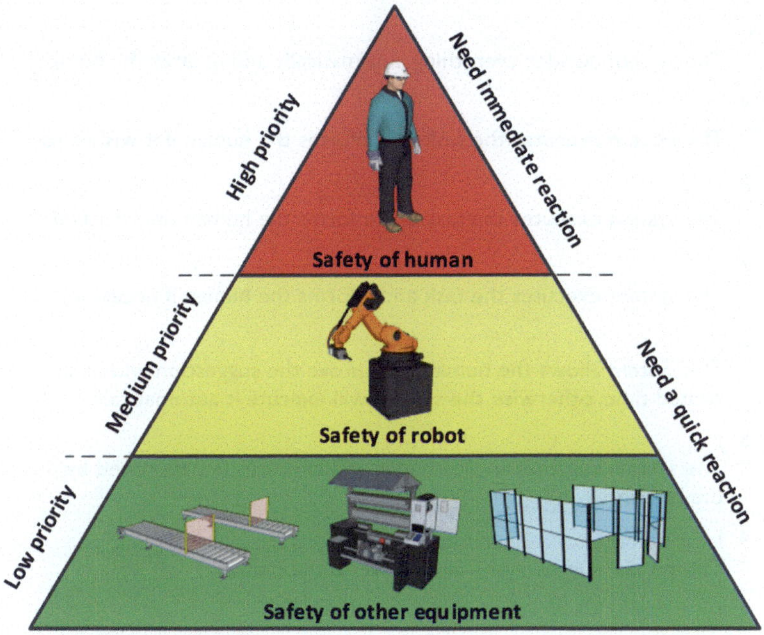

Fig. 1 Levels of safety in human-robot shared environment [5]

downtime and to avoid collisions caused by inaccurate programming. However, these tools require pre-knowledge about a robotic system, and any unpredicted changes in the robotic environment may cause a breakdown in the system since the robot program will no longer be valid.

Much research has focused on defining the framework for the remote collaboration between shop floor operators and industrial robots. One approach is that of Blech et al. [7] who presented a cyber-virtual system for modelling and visualizing remote industrial sites and showed the possibilities of imitating the behaviour of the physical robot using a virtual one. Another example is the research presented by Wang et al. [8] which focused on defining a distributed multi-robotic system that allows shop floor operators to control remote robots and perform industrial collaborative tasks. Results showed that their developed approach can be considered as a tool for training novice operators. Another approach reported by Junior [9] focused on developing a multilayer distributed architecture to control remote robots and to establish a collaborative behaviour among robots and humans driven by a set of defined rules. Other research such as that by Vartiainen et al. [10] indicated the possibility of establishing collaboration between a mobile robot and an expert operator. Research also showed that the approach can be used to perform maintenance and service tasks in industrial settings.

Other researchers focused on defining an appropriate approach to remote interaction between the human and the robot. For example, the work reported by Zhong

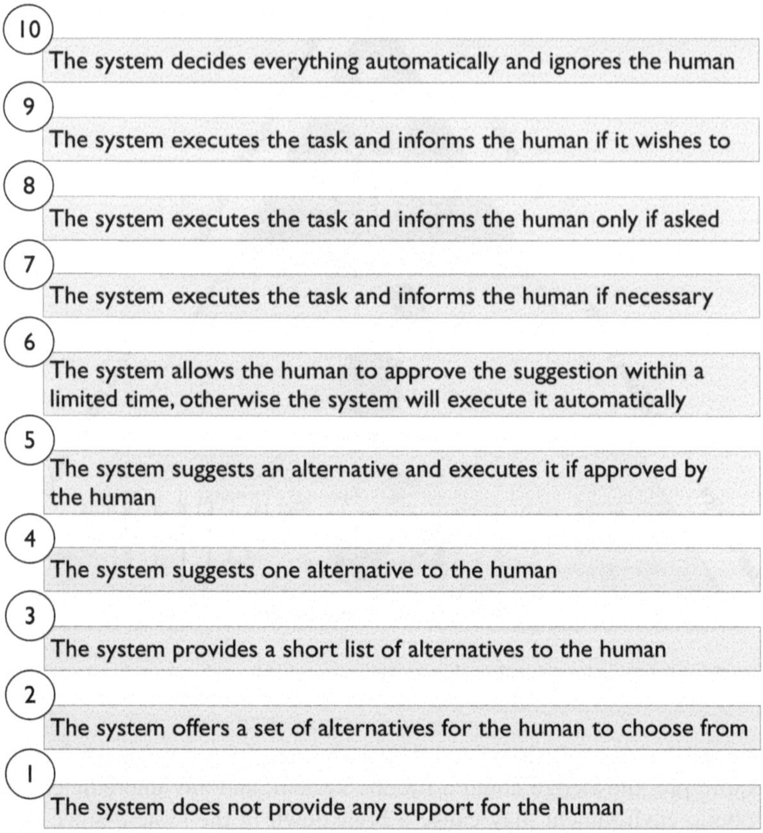

Fig. 2 Levels of automation

et al. [11] defined a cyber-hub infrastructure to allow the remote operators to control the robot using hand gestures. The developed approach showed that a group of distributed operators have the potential to simultaneously control several robots located in distant locations. Other researchers such as Itoh et al. [12] investigated the potential of using motion and force scaling to establish a human-robot remote collaborative manipulation based on virtual tool dynamics. Charles et al. [13] highlighted the potential of using a master-slave distributed system to remotely control a lightweight microsurgery robot. The operator in this case was assisted by feedback simulating the force reflections applied to the end-effector of the robot.

Establishing remote robotic laboratories was the main focus for other research groups such as Peake et al. [14]. This research focused on defining a platform for distant human-machine interaction using a simulation based cloud server for visualization. Another example is the research presented by Ashby [15] which developed a collaboration tool to allow trainees to log in to a remote laboratory which consists of robots and automation equipment.

Both laser scanners and vision cameras are common techniques to convert unknown objects to virtual 3D models. Modelling objects using stereo vision cameras was the main focus for several researchers including Karabegovic et al. [16]. The stereo vision camera-based approach suffers from two drawbacks:

(1) The equipment used is expensive and less compact

and

(2) Lacks the ability to capture and model complex shapes from a fixed single viewpoint due to limited visibility.

2D vision systems can also be applied to model unknown objects. By taking a number of snapshots of an object from different viewpoints, the object can be modelled based on the analysis of the silhouette in each snapshot. For example, Esteban and Schmitt [17] focused on modelling the object in high accuracy with details.

Despite the fact that these approaches were generally successful in their reported applications, they are unable to model multiple objects in a single run. Further, they lack the ability to model objects remotely.

Human-Robot Local Collaboration (HRLC)

A human-robot collaborative setting demands local cooperation between both humans and robots, and the safety of the humans in such a setting is an essential requirement. This is supported by both passive collision detection and active collision avoidance to track the shop floor operators and control the robots in a synchronized way.

In recent years, human-robot local collaboration has been reported by several researchers. Monje et al. [18] and Takata and Hirano [19] presented systems for controlling a humanoid robot working in cooperation with an operator. Takata and Hirano [19] also provided a tool to flexibly allocate shop floor operators and industrial robots in a shared assembly environment. Chen et al. [20] introduced a simulation based optimization approach with multiple objectives to define strategies for human-robot assembly operations. Krüger et al. [21] addressed the possibilities and advantages of employing multiple technologies in human-robot collaboration settings.

The main advantage of human-robot collaboration in a shared robotic environment is that of combining the high reliability of the robots with the high adaptability of the humans. However, such a setup when involving direct interaction needs a precise design to avoid any unnecessary stress for the humans working in that environment. Arai et al. [22] aimed to introduce a hybrid assembly setup by evaluating the stress level of a human working in close proximity to the robot.

Kuli and Croft [23] estimated the human's affective state in real-time by taking into consideration the robot movement as a stimulus and measuring human biological activity such as heart rate, facial expression and perspiration level. Meanwhile,

Charalambous et al. [24] reported a trust measurement tool that can be utilized for setting up industrial human-robot collaboration. This consisted of two parts; firstly, obtaining feedback from a number of operators using a defined survey, and secondly, using the participants' answers to a scenario involving three industrial robots. The first of these was a single-arm industrial robot having a payload of 45 kg, the second a twin-arm industrial robot having a payload of 20 kg and the third a single-arm industrial robot having a payload of 200 kg.

In a separate paper, Charalambous et al. [25] introduced guidelines to show the importance of human factors in the collaboration between the human and the robot. This research also explained how human factors can affect the level of success in establishing the collaboration. The human aspect is also highlighted by Ore et al. [26] who aimed to design an optimal human-robot collaborative environment. The approach considered operational time and biomechanical load to define methods of collaborations within industrial settings.

Charalambous et al. [27] also presented a framework to review the effect of the human organizational factors that need to be considered in order to have a successful human-robot collaborative setup. This was achieved by introducing a case study to evaluate the level of success for the collaboration between the human and the robot. The results of the study showed that several human factors, including participation, communication and management, can play important roles in the implementation of the human-robot collaboration. Elsewhere, Cherubini et al. [28] developed a sensor based framework that uses both depth and force sensors to control an industrial robot and establish a safe interaction between the human and the robot.

Other researchers focused on using different techniques to develop methods for human-robot collaborations. Ding et al. [29] used Finite State Automata (FSA) to develop an approach that considers the collaboration between multiple operators and multiple robots. Research by Geravand et al. [30] used physical interaction to define a collaboration between the human and the robot. Their presented approach measured the currents consumed by the robot's joints motors and controlled the robot accordingly to facilitate the collaboration. The approach relies on the fact that the current needed for each joint in the robot gives an indication of the external forces applied on those joints. Song et al. [31] introduced a collaborative robot that can interact with the human without the need for additional sensors by using the counter balanced mechanism (CBM). This mechanism uses springs to compensate for the torques required to operate the joints of the robot. Meanwhile, Michalos et al. [32] presented an augmented reality tool for supporting the human operator in a shared assembly environment. The tool uses video and text based visualizations to instruct the operator during the assembly process.

Other research aims to develop approaches that can perceive and protect humans working with robots in shared environments. These approaches are mainly based on two methods:

(1) The construction of a vision based system that can track the human in the robotic cell by combining the detection of the human's skin colour with a 3D model [33].

and

(2) An inertial sensor-based approach using a geometric representation of human operators generated with the help of a motion capture suit [34].

Real industrial experiments indicate that the latter approach may not be considered as a realistic solution as it relies heavily on the presence of a set of sensors attached to a uniform worn by the operator and the issues of capturing the movement around the person wearing the uniform, leaving neighbouring objects undetected. This can create a safety issue as there may be a possibility of collision between a moving object and a stationary operator. This is explained in detail together with other sensing methods in the literature survey of Bi and Wang [35].

The efficiency of vision based collision detection has been the interest of several research groups. For instance, Gecks and Henrich [36] developed a multi-camera collision detection system, whereas a high-speed emergency stop was utilized by Ebert et al. [37] to avoid a collision based on a specialized vision chip for tracking. Vogel et al. [38] presented a projector-camera based approach. This is achieved by identifying a protected zone around the robot and defining the boundary of that zone. The reported approach is capable of detecting dynamically and visually any safety interruption. Tan and Arai [39] presented an approach consisting of a triple stereovision system for capturing the motion of a seated operator (upper-body only) by wearing coloured markers. However, methods that depend on colour consistency may not be suitable in environments with uneven lighting conditions.

Markers for tracking moving operators may not be clearly visible within the monitored area. As an alternative to markers, a ToF (time-of-flight) camera was utilized by Schiavi et al. [40] for collision detection, and an approach using 3D depth information was introduced by Fischer and Henrich [41] for the same purpose. The use of laser scanners provides resolution but generates high computational latency, since each pixel or row of the captured scene needs to be analysed individually. Nonetheless, ToF cameras present a high performance solution for depth image acquisition, but with a low pixel resolution (with the ability to reach 200×200) and relatively high cost. Recently, Rybski et al. [42] built a three-dimensional grid to trace foreign objects and recognize humans, robots and background using data from 3D imaging sensors. Currently, Flacco et al. [43] use depth information from a single Kinect™ sensor to develop an approach for collision avoidance.

Other researchers aimed to integrate different sensing techniques to track humans and robots on shop floors, as in the work reported by Dániel et al. [44], by developing a collision-free robotic setting by combining the outputs of both ultrasonic and infrared proximity sensors. Researchers such as Cherubini et al. [45] introduced a direct interaction between the human and the robot by fusing both force/torque sensors and vision systems into a hybrid assembly setup.

Several commercial systems have also been presented as safety protection solutions. One of the widely accepted choices is SafetyEYE® [46], which uses a single stereo image to determine 2½D data of a monitored region and detect any interruption of predefined safety zones. An emergency stop will be triggered if an operator enters into any of the safety zones of the monitored environment. However, these safety zones are static and cannot be updated during run time.

In order to fulfil market demands for high productivity, several companies use a large number of industrial robots in their production lines. However, the same companies are today faced with new demands for a wide range of products with different specifications. To fulfil these new demands, many companies try to increase the adaptability of their production lines. This is a challenging task due to the fact that these production lines were designed initially for high productivity. One of the most effective solutions is to establish safe and collaborative environments in these production lines to allow the operators to work side by side with the robots.

It is important to maintain high productivity during human-robot collaboration. Therefore, there is a need for low cost and reliable online safety systems for effective production lines where shop floor operators share the tasks with industrial robots in a collision-free fenceless environment. Despite the fact that current safety approaches have been successful, the reported methods and systems are either highly expensive or relatively limited in handling real industrial applications. Further, most of the current commercial safety tools are based on defining static regions in the robotic cell with limited possibilities of close collaboration. Focusing on finding a solution, this part of the chapter presents a novel framework for introducing a safe and protected environment for human operators working locally with robots on the shop floor. These include:

(1) The successful recognition of any possible collision between the robot's 3D models and the human's point cloud captured by depth sensors in an augmented-reality environment.

and

(2) The active avoidance of any collision through active robot control. The developed system can dynamically represent the robot with a set of bounding boxes and move away the robot to avoid potential collision. This opens the door to several possibilities of closer collaborations between the human and the robot. Figures 3 and 4 explain this approach.

Human-Robot Collaboration Framework

This section presents the developed framework of this research, which can provide for and support collaborations between humans and robots. The first part describes the methodology and system implementation of remote human-robot collaboration, followed by the system implementation of local human-robot collaboration in the second part.

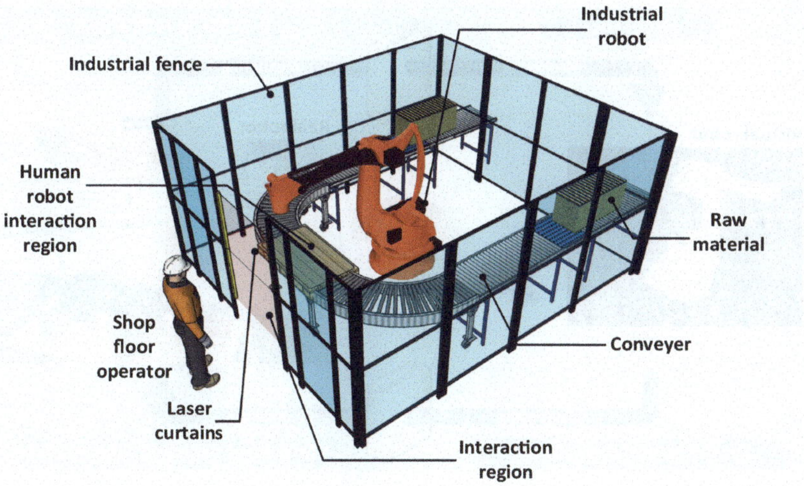

Fig. 3 Example of conventional industrial setup

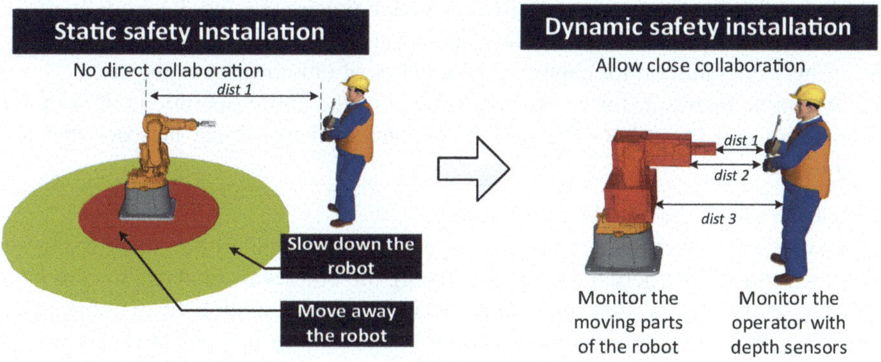

Fig. 4 From static to dynamic safety installations

Remote Human-Robot Collaboration

A new approach to simultaneously constructing 3D models of multiple arbitrary objects is proposed based on a set of snapshots taken of the objects from different angles. This approach is implemented through a system that allows an operator to perform assembly operations from a distance. Figure 5 illustrates the concept.

The proposed system demonstrates the ability to identify and model any unknown incoming objects to be assembled using an industrial robot [3]. The new objects are then integrated with the existing 3D model of the robotic cell in the structured environment of Wise-ShopFloor [6, 47], for 3D model-based remote assembly. The system consists of four modules:

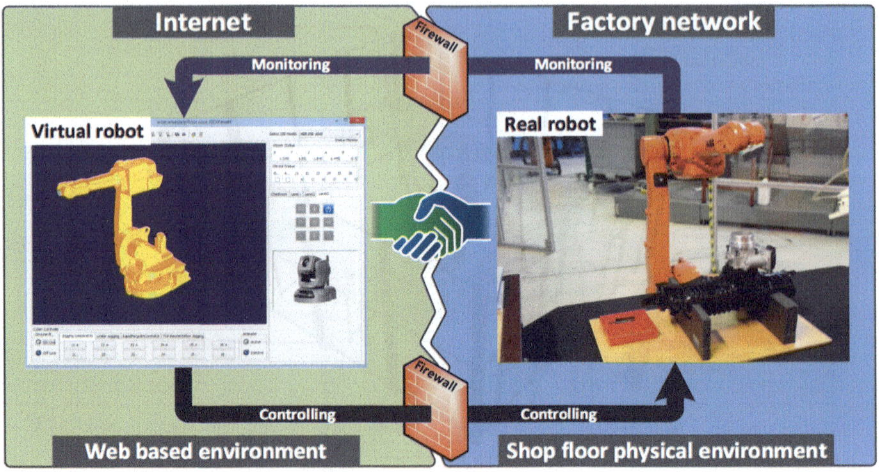

Fig. 5 Remote human-robot collaboration

(1) An application server, responsible for image processing and 3D modelling;
(2) A robot, for performing assembly operations;
(3) A network camera, for capturing silhouettes of unknown/new objects;
(4) A remote operator, for monitoring/control of the entire operations of both the camera and the robot. The system is connected to a local network and the Internet.

Figure 6 shows the details of the system.

The network camera is mounted on the end effector of the robot. First, the robot moves to a position where the camera is facing the objects from above to capture a top-view image. The system then constructs the primary models of the objects by converting their silhouettes in the top-view image to a set of vertical pillars with a default initial height. After that, the camera is used to take a sequence of images of the objects from other angles. Projecting the silhouettes of each image back to the 3D space generates a number of trimmed pillars. The intersections of these pillars identify the final 3D models of the objects.

A set of 10 experiments has been performed with the system to determine the average computation time for each processing step. Figure 7 then shows the percentage of the processing time required by each step with respect to the total time required to perform the image processing. It is found that although the system can process an image in ~4 s, some processing steps consumed a significant percentage of the total processing time. This is due to the fact that some image processing steps require examining each pixel of the captured images.

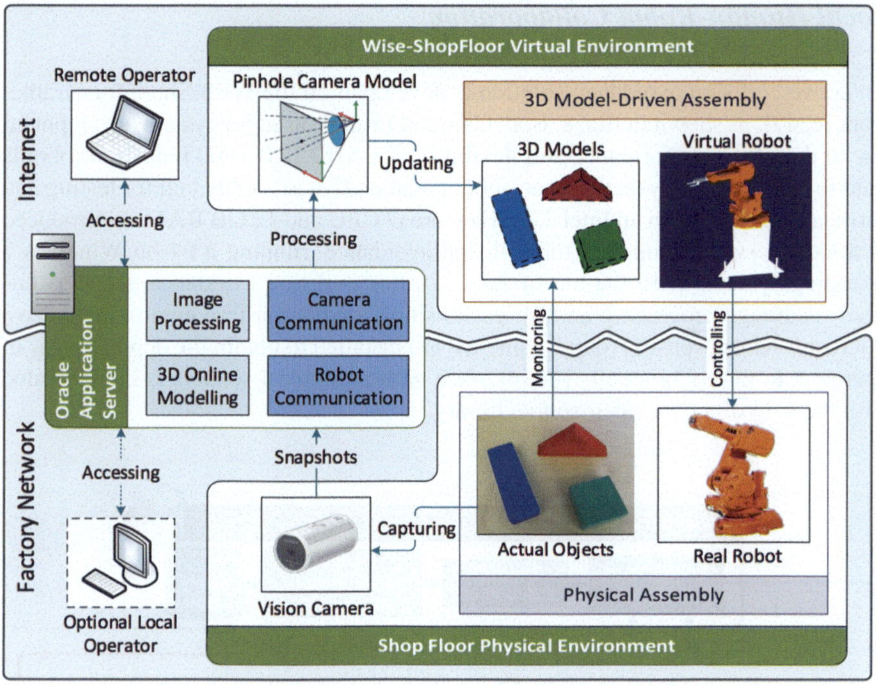

Fig. 6 System overview for remote human-robot collaborative assembly [3]

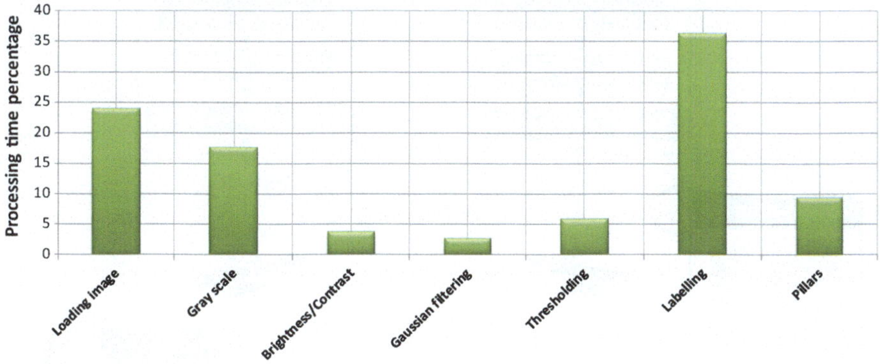

Fig. 7 Computation time for the processing steps of remote human-robot collaborative assembly [3]

Local Human-Robot Collaboration

An active collision avoidance solution is developed on the Wise-ShopFloor framework [6, 47], as shown in Fig. 8. Both C++ and Java are used for system development due to their high performance and flexibility. An ABB IRB 1600 industrial robot is used to construct a physical human-robot collaborative assembly cell for testing and verification. A PC with an Intel 2.7 GHz Core i7 CPU and 12 GB RAM is introduced as a local server responsible for collision avoidance, running a 64-bit Windows 7 operating system. With the aid of Java 3D, this collision avoidance server is utilized for image processing and for establishing a collision-free environment. Two Microsoft Kinect sensors (depth cameras) are installed to obtain the depth images of operators in the robotic cell. The interface with the Kinect sensors is implemented with the help of a C++ open-source library.

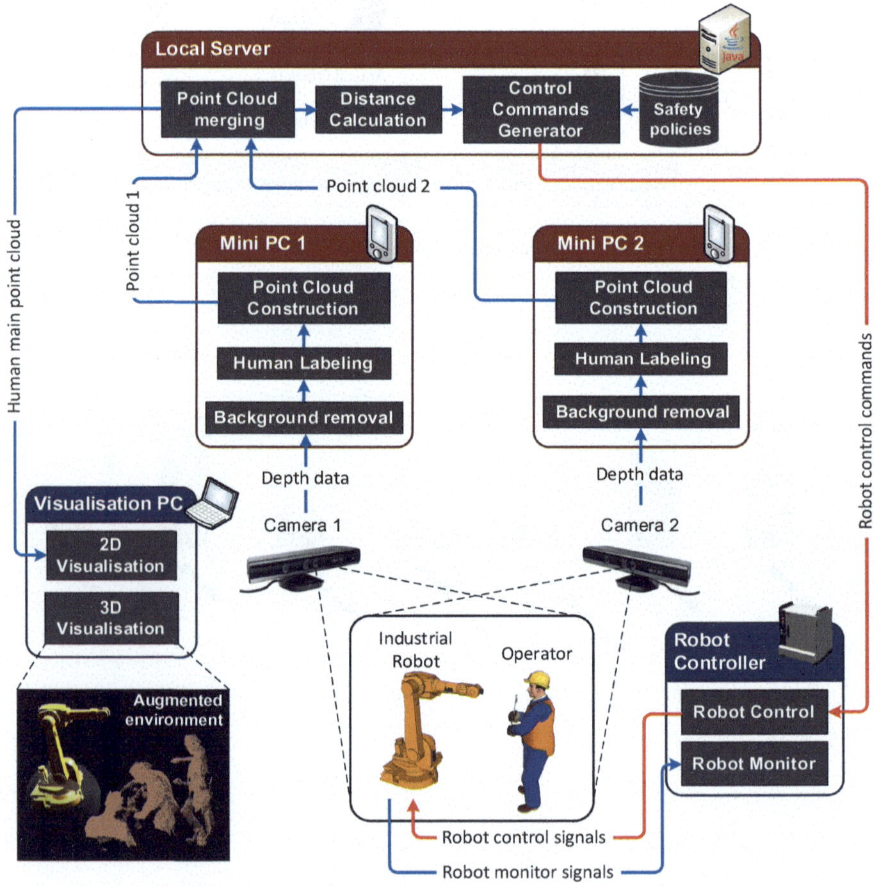

Fig. 8 Active collision avoidance system

The approach developed by Mohammed et al. [4] commences by calibrating the Kinect sensors and then acquiring the depth information from them. The process continues by determining the most proximate distance between the robot and obstacles (including operators) when active collision avoidance is performed. The velocity of the approaching operator is calculated to improve the system responsiveness by controlling the robot based on the velocity of the approaching human.

Additional experiments indicated that it is possible to determine the speed of any person approaching the robot in the robotic cell. The calculation of the human's speed is based on tracking the human's positions in 3D space in a number of multiple time periods. To evaluate the efficiency of the calculation, a configuration was established in which a pendulum was installed within the robotic cell and its movement tracked during steady-state swinging using the depth sensors.

The results of the experiment indicate that the system is able to detect the speed of the approaching human and control the robot accordingly. The results of this experiment then allow the developed system to predict the position of an obstacle and thus to handle the high-speed movements of the human and improve processing performance.

Discussions and Conclusions

The chapter focused on developing a framework for the creation of a safe human-robot collaborative production environment. It introduces two means of collaboration aimed at enhancing the sustainability of production lines through the introduction of innovative approaches and enhancement of the shop floor working environment. The results show that the systems are capable of achieving their intended objectives.

The first system provides an approach for remote operators to monitor and control an industrial robot in a virtual robotic environment. The 3D models of the components to be assembled are constructed based on images of the components captured by a robot-mounted IP camera. Due to the real-time network speed constraint, the camera cannot be used during assembly—thus this vision-assisted and 3D model-based approach. This allows the operator to perform remote assembly tasks using robots located in hazardous locations. A case study has been performed to evaluate the performance of the system. The results of the case study show that the developed system can generate a set of 3D models in about 24 s based on seven images. The efficiency can be improved by parallel image processing during the travel time when the robot is moving for the next image. Additional images can be used to improve the modelling accuracy of more complex objects. Overall, the accuracy of a 3D model may be attributed to:

(1) Camera calibration,
(2) Camera resolution,
(3) The distance between an object and the camera.

The second system is a tool for actively avoiding collisions between humans and robots, and which provides reliable and coherent safety protection in a human-robot collaborative environment. The aim of the system as developed is to enhance the overall performance of the robotic system. This is achieved by fusing the 3D models of the robots with the point clouds of the human operators in an augmented environment using a series of depth and vision sensors for active collision detection and avoidance. By comparing the developed approach with traditional safety systems, two major benefits can be identified; the first is that it allows for immersive collaboration between humans and robots and the second is the reduced downtime of the robotic cell when co-existing with humans. This approach can detect possible collisions in real-time and actively control the robot in one of four safety modes:

(1) Alerting the human operator.
(2) Stopping the robot.
(3) Moving the robot away from the approaching operator through recoverable interruptions.
(4) Online modification of the robot trajectory at runtime.

This approach thus provides better adaptability and productivity. Furthermore, a local human-robot collaborative scenario has been validated for the purpose of enabling the robot to track the operator to facilitate a shared assembly task.

A number of conditions are identified to improve the presented frameworks: firstly, the accuracy of the collaborative assembly tasks; secondly, the enhancement of the user interface to simplify the collaboration with the robot; thirdly, the definition of more industrial assembly cases for the developed systems and to analyse performance; and fourthly, to define and implement additional adaptive safety strategies to facilitate better collaboration between the human and the robot.

References

1. US Department of Commerce @ ecos.com/assets/uploads/2013/10/EFP-Sustainable-Manufacturing-Praciticesdited2016.pdf.
2. OECD. (2011). *OECD sustainable manufacturing tookit—Seven steps to environmental excellence* @ www.oecd.org/innovation/green/toolkit/48704993.pdf. Accessed May 20, 2019.
3. Wang, L., Mohammed, A., & Onori, M. (2014). Remote robotic assembly guided by 3D models linking to a real robot. *CIRP Annals—Manufacturing Technology, 63*(1), 1–4.
4. Mohammed, A., Schmidt, B., & Wang, L. (2016). Active collision avoidance for human–robot collaboration driven by vision sensors. *International Journal of Computer Integrated Manufacturing*, 1–11.
5. Holm, M., Givehchi, M., Mohammed, A., & Wang, L. (2012). Web based monitoring and control of distant robotic operations. In *ASME 2012 International Manufacturing Science and Engineering Conference collocated with the 40th North American Manufacturing Research Conference and in participation with the International Conference on Tribology Materials and Processing* (pp. 605–612).
6. Parasuraman, R., Sheridan, T. B., & Wickens, C. D. (2000). A model for types and levels of human interaction with automation. *IEEE Transactions on Systems, Man, and Cybernetics-Part A: Systems and Humans, 30*(3), 286–297.
7. Blech, J. O., Spichkova, M., Peake, I., & Schmidt, H. (2014). Cyber-virtual systems: Simulation, validation & visualization. In *9th International Conference on Evaluation of Novel Approaches to Software Engineering (ENASE)* (pp. 1–8).

8. Wang, X. G., Moallem, M., & Patel, R. V. (2003). An internet-based distributed multiple-telerobot system. *IEEE Transactions on Systems, Man, and Cybernetics-Part A: Systems and Humans, 33*(5), 627–633.
9. Junior, J. M., Junior, L. C., & Caurin, G. A. (2008). Scara3D: 3-Dimensional HRI integrated to a distributed control architecture for remote and cooperative actuation. In *Proceedings of 2008 ACM Symposium Applied Computing* (pp. 1597–1601).
10. Vartiainen, E., Domova, V., & Englund, M. (2015). Expert on wheels: An approach to remote collaboration. In *HAI 2015 Proceedings of the 3rd International Conference on Human-Agent Interaction* (pp. 49–54).
11. Zhong, H., Wachs, J. P., & Nof, S. Y. (2013). HUB-CI model for collaborative telerobotics in manufacturing. *IFAC Proceedings, 46*(7), 63–68.
12. Itoh, T., Kosuge, K., & Fukuda, T. (2000). Human-machine cooperative telemanipulation with motion and force scaling using task-oriented virtual tool dynamics. *IEEE Transactions on Robotics and Automation, 16*(5), 505–516.
13. Charles, S., Das, H., Ohm, T., Boswell, C., Rodriguez, G., Steele, R., & Istrate, D. (1997). Dexterity-enhanced telerobotic microsurgery. In *Proceedings of the 8th International Conference on Advanced Robotics ICAR'97* (pp. 5–10).
14. Peake, I., Blech, J. O., Fernando, L., Schmidt, H., Sreenivasamurthy, R., & Sudarsan, S. D. (2015). Visualization facilities for distributed and remote industrial automation: VxLab. In *Proceedings of the IEEE International Conference on Emerging Technologies & Factory Automation* (pp. 1–4).
15. Ashby, J. E. (2008). The effectiveness of collaborative technologies in remote lab delivery systems. In *Proceedings of the 38th Frontiers of Education Conference* (pp. 7–12).
16. Karabegovic, I., Vojic, S., & Dolecek, V. (2006). 3D vision in industrial robot working process. In *12th International Power Electronics and Motion Control Conference* (pp. 1223–1226).
17. Esteban, C. H., & Schmitt, F. (2004). Silhouette and stereo fusion for 3D object modeling. *Computer Vision and Image Understanding, 96*(3), 367–392.
18. Monje, C. A., Pierro, P., & Balaguer, C. (2011). A new approach on human–robot collaboration with humanoid robot RH-2. *Robotica, 29*(6), 949–957.
19. Takata, S., & Hirano, T. (2011). Human and robot allocation method for hybrid assembly systems. *CIRP Annals—Manufacturing Technology, 60*(1), 9–12.
20. Chen, F., Sekiyama, K., Huang, J., Sun, B., Sasaki, H., & Fukuda, T. (2011). An assembly strategy scheduling method for human and robot coordinated cell manufacturing. *International Journal of Intelligent Computing and Cybernetics*, 487–510.
21. Krüger, J., Lien, T. K., & Verl, A. (2009). Cooperation of human and machines in assembly lines. *CIRP Annals—Manufacturing Technology, 58*(2), 628–646.
22. Arai, T., Kato, R., & Fujita, M. (2010). Assessment of operator stress induced by robot collaboration in assembly. *CIRP Annals—Manufacturing Technology, 59*(1), 5–8.
23. Kuli, D., & Croft, E. A. (2007). Affective state estimation for human–robot interaction. *IEEE Transactions on Robotics, 23*(5), 991–1000.
24. Charalambous, G., Fletcher, S., & Webb, P. (2016). The development of a scale to evaluate trust in industrial human-robot collaboration. *International Journal of Social Robotics, 8*(2), 193–209.
25. Charalambous, G., Fletcher, S., & Webb, P. (2016). Development of a human factors roadmap for the successful implementation of industrial human-robot collaboration. In *Proceedings of the AHFE 2016 International Conference on Human Aspects of Advanced Manufacturing* (pp. 195–206).
26. Ore, F., Vemula, B. R., Hanson, L., & Wiktorsson, M. (2016). Human–industrial robot collaboration: Application of simulation software for workstation optimisation. *Procedia CIRP, 44*, 181–186.
27. Charalambous, G., Fletcher, S., & Webb, P. (2015). Identifying the key organisational human factors for introducing human-robot collaboration in industry: an exploratory study. *The International Journal of Advanced Manufacturing Technology, 81*(9–12), 2143–2155.

28. Cherubini, A., Passama, R., Fraisse, P., & Crosnier, A. (2015). A unified multimodal control framework for human-robot interaction. *Robtics & Autonomous Systems, 70,* 106–115.
29. Ding, H., Schipper, M., & Matthias, B. (2013). Collaborative behavior design of industrial robots for multiple human-robot collaboration. In *IEEE 44th International Symposium on Robotics (ISR 2013)* (Vol. 49, pp. 1–6).
30. Geravand, M., Flacco, F., & De Luca, A. (2013). Human-robot physical interaction and collaboration using an industrial robot with a closed control architecture. In *IEEE International Conference on Robotics and Automation (ICRA)* (pp. 4000–4007).
31. Song, S.-W., Lee, S.-D., & Song, J.-B. (2015). 5 DOF industrial robot arm for safe human-robot collaboration. In *8th International Conference on Intelligent Robotics & Applications (ICIRA 2015)* (pp. 121–129).
32. Michalos, G., Karagiannis, P., Makris, S., Tokçalar, Ö., & Chryssolouris, G. (2016). Augmented Reality (AR) applications for supporting human-robot interactive cooperation. *Procedia CIRP, 41,* 370–375.
33. Krüger, J., Nickolay, B., Heyer, P., & Seliger, G. (2005). Image based 3D surveillance for flexible man-robot-cooperation. *CIRP Annals—Manufacturing Technology, 54*(1), 19–22.
34. Corrales, J. A., Candelas, F. A., & Torres, F. (2011). Safe human-robot interaction based on dynamic sphere-swept line bounding volumes. *Robotics & Computer Integrated Manufacturing, 27*(1), 177–185.
35. Bi, Z. M., & Wang, L. (2010). Advances in 3D data acquisition and processing for industrial applications. *Robotics & Computer Integrated Manufacturing, 26*(5), 403–413.
36. Gecks, T., & Henrich, D. (2005). Human-robot cooperation: Safe pick-and-place operations. In *Proceedings of the IEEE International Workshop on Robot & Human Interactive Communication (ROMAN 2005)* (pp. 549–554).
37. Ebert, D., Komuro, T., Namiki, A., & Ishikawa, M. (2005). Safe human-robot-coexistence: Emergency-stop using a high-speed vision-chip. In *IEEE/RSJ International Conference on Intelligent Robotic Systems (IROS 2005)* (pp. 1821–1826).
38. Vogel, C., Walter, C., & Elkmann, N. (2013). A projection-based sensor system for safe physical human-robot collaboration. In *IEEE International Conference on Intelligent Robotic Systems (IROS 2013)* (pp. 5359–5364).
39. Tan, J. T. C., & Arai, T. (2011). Triple stereo vision system for safety monitoring of human-robot collaboration in cellular manufacturing. In *Proceedings of the IEEE International Symposium on Assembly & Manufacturing (ISAM 2011)* (pp. 1–6).
40. Schiavi, R., Bicchi, A., & Flacco, F. (2009). Integration of active and passive compliance control for safe human-robot coexistence. In *Proceedings of the IEEE International Conference on Robot Automation* (pp. 259–264).
41. Fischer, M., & Henrich, D. (2009). 3D collision detection for industrial robots and unknown obstacles using multiple depth images. *Advances in Robotics Research,* 111–122.
42. Rybski, P., Anderson-Sprecher, P., Huber, D., Niessl, C., & Simmons, R. (2012). Sensor fusion for human safety in industrial workcells. In *Proceedings of the IEEE International Conference on Intelligent Robotic Systems* (pp. 3612–3619).
43. Flacco, F., Kroeger, T., De Luca, A., & Khatib, O. (2015). A Depth space approach for evaluating distance to objects: with application to human-robot collision avoidance. *Journal of Intelligent & Robotic Systems Theory, 80,* 7–22.
44. Dániel, B., Korondi, P., & Thomessen, T. (2012). Joint level collision avoidance for industrial robots. *Proceedings of the IFAC, 45*(22), 655–658.
45. Cherubini, A., Passama, R., Crosnier, A., Lasnier, A., & Fraisse, P. (2016). Collaborative manufacturing with physical human-robot interaction. *Robotics & Computer Integrated Manufacture, 40,* 1–13.
46. Pilz GmbH & Co. KG @ www.safetyeye.com/. Accessed May 20, 2019.
47. Wang, L. (2008). Wise-ShopFloor: An integrated approach for web-based collaborative manufacturing. *IEEE Transactions on Systems, Man, and Cybernetics, Part C (Applications and Reviews), 38*(4), 562–573.

Tracking Control for a Fully-Actuated UAV

Matthias Konz, David Kastelan and Joachim Rudolph

Abstract The realization of an unmanned aerial vehicle (UAV) with three tiltable propellers in a planar 120° arrangement is described in detail. A single rigid body approximate model of this vehicle is fully actuated, thus allowing for a variety of flight maneuvers including translation without tilting and inclined hovering. The tracking control of this inherently unstable system is treated in existing literature but assumes full-state measurement and direct access to the body-fixed forces and torques. To meet these challenges, a fast control scheme for the underlying actuator dynamics is proposed and a state and disturbance estimator based on on-board inertial and infrequent delayed external position and orientation measurements is discussed. All designs are experimentally validated including flight tests that demonstrate good trajectory tracking performance.

Introduction

Small unmanned aerial vehicles (UAVs) continue to be a topic of significant research interest, due to their proven versatility as sensing and transport platforms. Recent reviews of applications along with related opportunities and challenges are provided by Floreano and Wood [1], Kumar and Michael [2]. For indoor and short-range applications, vehicle designs have all but converged on the so-called multicopter arrangement of direct-drive fixed-pitch propellers that provide the forces and torques necessary for stabilization and navigation.

Typically, these actuators are introduced pairwise, such that inherent aerodynamic reaction torques cancel in the nominal hover case through the use of counter-rotating propellers. Furthermore, their orientations with respect to the vehicle frame are usually fixed in the vertical direction, as to maximize weight-countering thrust in hover. The results are mechanically simple and robust vehicles with largely decoupled dynamics that may be effectively treated by even elementary control methods. Given their large thrust to weight ratios and a variety of control strategies, many interesting

M. Konz (✉) · D. Kastelan · J. Rudolph
Saarland University, Saarbrücken, Germany
e-mail: m.konz@lsr.uni-saarland.de

X.-T. Yan et al. (eds.), *Reinventing Mechatronics*,
https://doi.org/10.1007/978-3-030-29131-0_9

examples of body-fixed thrust vehicle capabilities have been demonstrated, as for instance by Mellinger and Kumar [3].

Despite their ubiquity, the manoeuvrability of these body-fixed thrust vehicles is fundamentally limited by the magnitude of the propeller reaction torque and the fact that only four of the vehicle's six degrees-of-freedom are actuated. Both of these issues may be addressed by actively directing the thrust with respect to the vehicle frame. In doing so, the pairwise arrangement of propellers may also be relaxed. The most straightforward implementation of this concept is the tricopter with three propellers, one of which may be tilted from vertical by way of a servo motor. Such a vehicle is described by Salazar-Cruz, Lozano and Escareño [4] and is well-known among multi-copter hobbyists, since its four actuators may be manually controlled by a pilot with a standard two-joystick (i.e. four-channel) remote control as in the body-fixed thrust vehicle case. The tilting action may be used to cancel the nominal non-zero reaction torque resulting from the odd number of propellers and, at the same time, enable comparatively high body yaw rates. Propeller tilting may also be used to directly influence vehicle degrees of freedom that are not actuated in a conventional body-fixed thrust multi-copter. This approach is taken by Ryll et al. [5] where a quadrotor is described with four servos that independently tilt each of the four propellers. The arrangement of these eight actuators results in an ambiguous relationship between their configuration and the actuation of the vehicle's six degrees-of-freedom. An online energy-minimizing optimization strategy is then used to uniquely determine the actuator configuration.

The present work describes a simpler, yet just as capable, vehicle with three independent tilting propellers. Their arrangement is chosen such that the six degrees-of-freedom of a single rigid body model of the vehicle is fully actuated with respect to the six vehicle inputs. These are assumed to be the three arm servo angles and the three propeller thrusts. In addition, a planar 120° arm arrangement is chosen as to make lateral and longitudinal vehicle motions equally favourable. Ignoring the practical caveat of actuator saturation, the result is a vehicle capable of flying arbitrary trajectories. These include useful manoeuvres unachievable by conventional body-fixed thrust multi-copters such as horizontal translation without tilting or hovering at an angle. The platform may likewise be used as a general purpose force and torque effector in three-dimensional space.

The LSR tricopter design and a pilot-supporting control were first discussed by Kastelan et al. [6]. In this chapter a full configuration (position and attitude) tracking control for the tricopter is presented. Since the theory for this control is well-established in the dedicated literature, the focus here lies rather on two main practical challenges of its implementation: Firstly, the real-time estimation of the configuration and velocity state of the tricopter as well as the treatment of unmodeled disturbances; and secondly, the fast and accurate realization of computed control forces by the vehicle's actuators. The performance of the complete implementation is demonstrated in experimental results for tracking of configuration transitions.

Realization

The developed tricopter shown in Fig. 1 is characterized by the three independently tiltable, propeller supporting arms. The central body and the arms are a sandwich construction of carbon-fibre plates and tubes and 3D-printed parts. The outer carbon-fibre rings serve as collision protection and landing gear. The vehicle has an outer diameter of about 0.8 m and a mass of about 1.2 kg. The tilting mechanism is driven by a standard hobby servo motor that is connected to a gear on the arm by a toothed-belt and allows for tilt angles of $\pm 75°$. The three 10 inch (25.4 cm) propellers are each driven by a brushless dc-motor (BLDC) capable of angular velocities up to 120 Hz corresponding to a maximum thrust of 8 N. As energy source, the tricopter carries a 14.8 V, 2.2 Ah, LiPo battery that allows for up to 15 min of autonomous flight.

Two types of sensors are used by the tricopter. An inertial measurement unit (IMU) VN-100 from VectorNav is mounted on the central body to measure its angular velocity and acceleration and provide attitude estimates. An external camera-based motion capture system from Vicon measures position and attitude by way of reflective markers on the tricopter.

A standard two-joystick remote control is used as a reliable and near real-time user interface to the vehicle while a pair of XBee S6B Wi-Fi modules provides a two-way communication link with a ground station PC. The important hardware components are summarized in Fig. 2.

The on-board electronics consist mainly of a custom-built mainboard and three identical motor drivers. The mainboard contains an Atmel AT32UC3C 32-bit, 66 MHz microcontroller with FPU and is tasked with executing all control algorithms. Although compact brushless-dc motor drivers are ubiquitous, available products typically lack explicit speed control. Moreover, while the rotor speed is implicitly known to the driver through electrical commutation, speed feedback necessary for

Fig. 1 The LSR tricopter

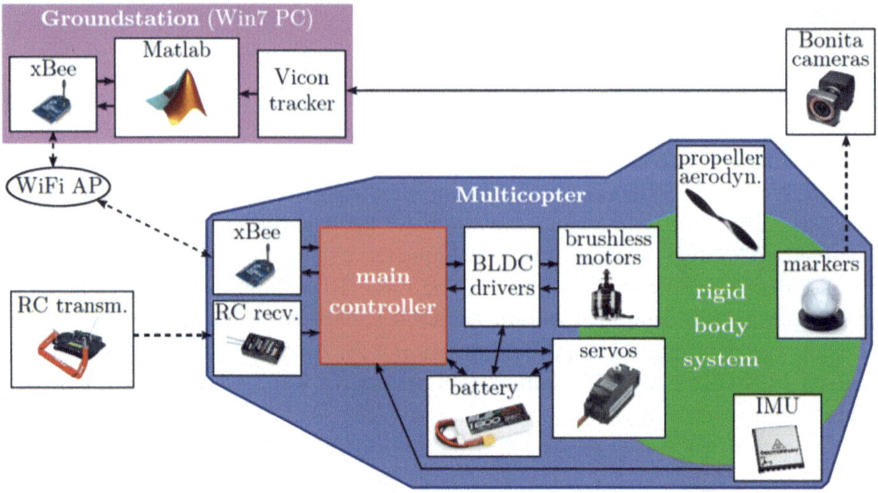

Fig. 2 Realization overview

an external control implementation is usually not provided. As a result, a custom module was developed that provides such a measurement while implementing an underlying current controller to drive the motor. This solution allows for rapid propeller dynamics, described later in the chapter, that includes active braking, achieved by feeding current back into the battery.

Conception, realization and validation of the presented tricopter involve the disciplines of mechanical, electrical and software engineering. Since issues in these fields are strongly coupled, they must to be tackled simultaneously and systematically. Thus the tricopter may be regarded as a classical example of a *mechatronic system*.

Mathematical Model

Rigid Body Dynamics

Mechanically, the tricopter is a system of 10 rigid bodies: a central body, 3 arms, 3 servo motor rotors, and 3 propeller assemblies, see Fig. 3. While the servo rotors are very small, since they sit behind a gearbox with ratio $c = 625$, their inertia that contributes to the kinetic equation with a factor c^2 is non-negligible. Since the propeller rotation rates are generally much faster (60–120 Hz) than any other dynamics of the tricopter, their inertia is modelled as a disc symmetric about their rotational axes, which is consequently independent of the propeller angles.

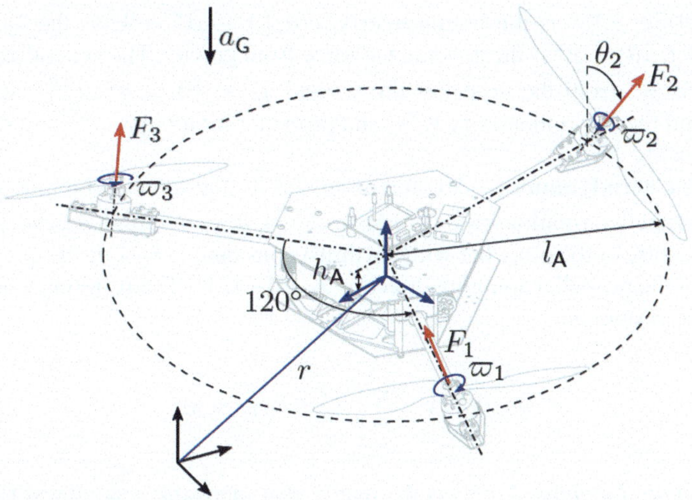

Fig. 3 Tricopter geometry

The configuration of the rigid body system is parameterized by the position $r \in \mathbb{R}^3$ of the central body, its orientation matrix $R \in SO(3)$, and the arm tilt angles $\theta \in \mathbb{R}^3$. For the velocity state of the system, the body-fixed velocity $v \in \mathbb{R}^3$ of the center body, its angular velocity $\omega \in \mathbb{R}^3$, the arm tilt velocities $\dot{\theta} \in \mathbb{R}^3$, and the propeller angular velocities $\varpi \in \mathbb{R}^3$ are chosen. These are collected in the generalized velocity vector $\xi = \left[v^T, \omega^T, \dot{\theta}^T, \varpi^T\right]^T \in \mathbb{R}^{12}$ and are related to the configuration coordinates through

$$\dot{r} = Rv, \quad \dot{R} = R\text{wed}(\omega) \tag{1}$$

where the *wedge* and *vee* operators are defined as

$$\text{wed}\begin{bmatrix} a_1 \\ a_2 \\ a_3 \end{bmatrix} = \begin{bmatrix} 0 & -a_3 & a_2 \\ a_3 & 0 & -a_1 \\ -a_2 & a_1 & 0 \end{bmatrix}, \ \text{vee}\begin{bmatrix} * & A_{12} & A_{13} \\ A_{21} & * & A_{23} \\ A_{31} & A_{32} & * \end{bmatrix} = \frac{1}{2}\begin{bmatrix} A_{32} - A_{23} \\ A_{13} - A_{31} \\ A_{21} - A_{12} \end{bmatrix}.$$

The kinetic energy $T(\theta, \xi) = \frac{1}{2}\xi^T M(\theta)\xi$ and potential energy $V_G(r, R, \theta)$ from gravity may now be formulated in terms of the chosen coordinates in order to derive the equations of motion. To this end, a formalism that handles this parameterization that includes redundant configuration coordinates and nonholonomic velocity coordinates ξ as proposed by Konz and Rudolph [7] is used. The resulting $n = 12$ kinetic equations take the form

$$M(\theta)\dot{\xi} + c(\theta, \xi) + f_G(\theta) = f_A(\theta, \varpi) + f_u(\tau_S, \tau_P), \tag{2}$$

where $M(\theta) \in \mathbb{R}^{n \times n}$ is the inertia matrix, $c(\theta, \xi) \in \mathbb{R}^n$ collects the gyroscopic forces, and $f_G(\theta) \in \mathbb{R}^n$ is the generalized force from gravity. The generalized forces $f_A(\theta, \varpi) \in \mathbb{R}^n$ from the propeller thrusts and $f_u = \left[0_{1\times 6}, \tau_S^T, \tau_P^T\right]^T$ due to the torques from the servo motors $\tau_S \in \mathbb{R}^3$ and propeller motors $\tau_P \in \mathbb{R}^3$ are discussed in the following.

Propeller aerodynamics—A common model for the force and torque resulting from the propeller rotation ϖ_i is $f_{P_i}(\varpi_i) = \left[0, 0, \kappa_F \varpi_i^2, 0, 0, \varepsilon_i \kappa_\tau \varpi_i^2\right]^T$ at the geometric centre of the propeller with the thrust constant $\kappa_F > 0$, the torque constant $\kappa_\tau > 0$, and the propeller spinning direction $\varepsilon_i \in \{-1, 1\}$. The resulting generalized force on the system is

$$f_A(\theta, \varpi) = \sum_{i=1}^{3} J_{P_i}^T(\theta_i) f_{P_i}(\varpi_i),$$

where the *body Jacobian* $J_{P_i}(\theta_i)$ is the matrix that maps the generalized velocity ξ to the absolute velocity of the corresponding rigid body, as for instance Murray et al. [8].

Servo motor—The servo includes an integrated controller assumed to be of the form

$$\tau_{S_i} = K_P\left(\theta_{R,i} - \theta_i\right) - K_D \dot{\theta}_i, \quad i = 1, 2, 3$$

where $\theta_{R,i}$ are the desired servo angles. Experiments validate this model as demonstrated later in Fig. 4.

Propeller motor—As discussed earlier, the propeller motor is current-controlled, thus resulting in negligible electrical dynamics on the time-scale of this mechanical model. Consequently, the motor currents $j_{\mathrm{M,i}}$ are considered as control inputs. Assuming some constant dry friction τ_0 on the motor shaft, the propeller motor torque is

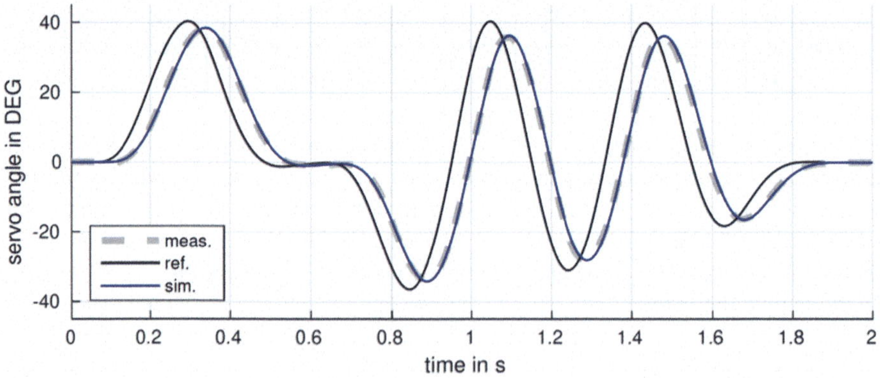

Fig. 4 Measured servo response and simulation

$$\tau_{P,i} = k_M j_{M,i} - \tau_0, \quad i = 1, 2, 3$$

Simplified Cascade Model

During the design of the tricopter, care was taken to reduce the effect of propeller tilting on the overall centre of mass and inertia (the central body accounts for approx. 75% of the total mass and the combined centre of mass of the arm and propeller bodies is very close to the tilting axis). As a result, these quantities may be assumed configuration-independent. By neglecting some further weak inertial coupling, the kinetic Eq. (2) simplifies significantly. At the same time, an additional force F_B and torque τ_B are added to account for model uncertainties and external disturbances. The result is the simplified tricopter model suitable for control design and consisting of the following parts:

1. Single rigid body dynamics with constant disturbance

$$m(\ddot{r} - a_G) = R(F_A + F_B), \quad \dot{F}_B = 0 \tag{3a}$$

$$\dot{R} = R\text{wed}(\omega), \quad \Theta\dot{\omega} + \text{wed}(\omega)\Theta\omega = \tau_A + \tau_B, \quad \dot{\tau}_B = 0 \tag{3b}$$

2. Transformation of the propeller force (with $\sigma_i = \varepsilon_i \kappa_\tau / \kappa_F$)

$$\begin{bmatrix} F_{A1} \\ F_{A2} \\ F_{A3} \\ \tau_{A1} \\ \tau_{A2} \\ \tau_{A3} \end{bmatrix} = \begin{bmatrix} 0 & 0 & 0 & \frac{\sqrt{3}}{2} & 0 & \frac{\sqrt{3}}{2} \\ 0 & 0 & 0 & -\frac{1}{2} & 1 & -\frac{1}{2} \\ 1 & 1 & 1 & 0 & 0 & 0 \\ \frac{\sqrt{3}l_A}{2} & 0 & -\frac{\sqrt{3}l_A}{2} & \frac{h_A - \sqrt{3}\sigma_1}{2} & h_A & \frac{h_A + \sqrt{3}\sigma_3}{2} \\ -\frac{l_A}{2} & l_A & -\frac{l_A}{2} & \frac{\sqrt{3}h_A + \sigma_1}{2} & \sigma_2 & \frac{\sigma_3 - \sqrt{3}h_A}{2} \\ -\sigma_1 & -\sigma_2 & -\sigma_3 & -l_A & -l_A & -l_A \end{bmatrix} \begin{bmatrix} F_1 \cos\theta_1 \\ F_2 \cos\theta_2 \\ F_3 \cos\theta_3 \\ F_1 \sin\theta_1 \\ F_2 \sin\theta_2 \\ F_3 \sin\theta_3 \end{bmatrix} \tag{3c}$$

3. Servo dynamics

$$\ddot{\theta}_i + 2\zeta_S \lambda_S \dot{\theta}_i + \lambda_S^2 \theta_i = \lambda_S^2 \theta_{R,i}, \quad i = 1, 2, 3 \tag{3d}$$

4. Propeller dynamics

$$F_i = \kappa_F \varpi_i^2, \quad \Theta\dot{\varpi}_i + \kappa_\tau \varpi_i^2 = k_M j_{M,i} - \tau_0, \quad i = 1, 2, 3 \tag{3e}$$

The parameters for this model, summarized in Table 1, were either directly measured or identified in dedicated experiments. Note that for the given parameters the relationship of Eq. (3c) is invertible, i.e. for a given force and torque (F_A, τ_A) it is possible to compute the corresponding propeller thrusts F_i and servo angles θ_i,

Table 1 Simplified tricopter model parameters

$\mathrm{m} = 1.245\,\mathrm{kg}$, $\Theta = \mathrm{diag}(18, 18, 27) \times 10^{-3}\,\mathrm{kg\ m^2}$, $a_G = [0, 0, -9.81]^T\ \mathrm{m/s^2}$,
$l_A = 0.24\,\mathrm{m}$, $h_A = 0.005\,\mathrm{m}$, $\varepsilon_1 = -1$, $\varepsilon_2 = 1$, $\varepsilon_3 = 1$,
$\Theta_P = 3.57 \times 10^{-5}\,\mathrm{kg\ m^2}$, $\kappa_F = 1.44 \times 10^{-5}\,\mathrm{N\ s^2}$, $\kappa_\tau = 2.24 \times 10^{-7}\,\mathrm{Nm\ s^2}$,
$\lambda_S = 39\,\mathrm{s}^{-1}$, $\zeta_S = 0.93$

$i = 1, 2, 3$. The rigid body model is therefore fully-actuated, such that the translational and rotational dynamics may be explicitly (and independently) prescribed.

Rigid Body Control

Tracking control for fully-actuated mechanical systems is well-established in the dedicated literature, e.g. [8]. Here, linear dynamics are used for the position error $r_E = r - r_R$ and the nonlinear dynamics proposed by Koditschek [9] for the attitude error $R_E = R_R^T R$ and its velocity $\omega_E = \omega - R_E^T \omega_R$

$$m\ddot{r}_E + K_v \dot{r}_E + K_r r_E = 0, \tag{4a}$$

$$\Theta\dot{\omega}_E + \mathrm{wed}(\omega_E)\Theta\omega_E + K_\omega\omega_E + 2\mathrm{vee}\Big(K_R^{'} R_E\Big) = 0. \tag{4b}$$

The explicit control law results from combining the desired error dynamics (4) with the rigid body model Eq. (3a–3c) to find

$$F_A = R^T(m(\ddot{r}_R - a_G) - K_v\dot{r}_E - K_r r_E) - F_B, \tag{5a}$$

$$\begin{aligned}\tau_A = \Theta\big(R_E^T\dot{\omega}_R - \mathrm{wed}(\omega_E)R_E^T\omega_R\big) + \mathrm{wed}(\omega_E)\Theta\omega_E - \mathrm{wed}(\omega)\Theta\omega \\ - K_\omega\omega_E - 2\mathrm{vee}\Big(K_R^{'} R_E\Big) - \tau_B.\end{aligned} \tag{5b}$$

Tuning—In contrast to the position error dynamics (4a), the attitude error dynamics of Eq. (4b) are less intuitive. To resolve this consider its *small angle approximation* $R_E \approx I_3 + \mathrm{wed}(\varepsilon)$ where $\varepsilon \in \mathbb{R}^3$ can be thought of as roll, pitch, and yaw angles or the vector part of the corresponding attitude quaternion. The approximation of the error dynamics, Eq. (4b), is

$$\Theta\ddot{\varepsilon} + K_\omega\dot{\varepsilon} + K_R\varepsilon = 0.$$

The matrices $K_R, K_\omega \in \mathbb{R}^{3\times 3}$ should be intuitive to tune for an engineer and the actual gain K_R' used in Eqs. (4b) and (5b) is calculated from $K_R' = \frac{1}{2}\mathrm{tr}(K_R)I_3 - K_R$.

Realization—The control law of Eq. (5a) assumes that there is direct control over F_A and τ_A, which is not the case since they are subject to the actuator dynamics of Eqs. (3c–3e). Furthermore it assumes perfect knowledge of the rigid body configuration (r, R), its velocity (v, ω), and the disturbances (F_B, τ_B), of which only the angular velocity ω is directly measured. A main subject of this chapter and the content of the next two sections is the realization of these control inputs and the estimation of the model state.

Another minor detail arises from the discrete-time (sampling time $T_S = 5$ ms) implementation: the computed controls cannot be applied immediately, but rather at the next sampling step, i.e. $j_M[k+1]$ and $\theta_R[k+1]$ are the available control inputs. To account for this, compute e.g. $F_A[k+1]$ from the predictions $r[k+1]$ etc. that are available from the implemented observers.

Actuator Dynamics and Control

For a given desired body-fixed force $F_A[k+1]$ and torque $\tau_A[k+1]$, it is possible to compute each required propeller thrust $F_{R,i}[k+1]$ and servo angle $\theta_{R,i}[k+1]$ by inverting Eq. (3c). Since the three servos and propellers are identical, the index i for the corresponding quantities is dropped in the following section for readability.

Servo Dynamics

The integrated servo controller was tested to be quite fast and accurate, though the internal angle measurement is not available. However, an online simulation using the forward Euler discretization of the servo dynamics of Eq. (3d) was shown to yield sufficiently good estimates $\hat{\theta}$ of the servo angle. These results are shown in Fig. 4 and compared with an external encoder measurement.

Thrust Dynamics

Thrust observer—The velocity measurement ϖ_M is noisy and therefore not suitable for direct feedback. Since a simple low-pass filtering would lead to an undesirable delay, a better solution is an observer. Its implementation is a copy of the (forward Euler) discretization of the nonlinear model of Eq. (3e) supplemented by a linear feedback proportional with gain l_ϖ to the error w.r.t. the measurement, i.e.

$$\hat{F}[k] = \kappa_F \left(\hat{\varpi}[k]\right)^2 \tag{6}$$

$$\hat{\varpi}[k+1] = \hat{\varpi}[k] + \frac{T_S}{\Theta_P}\Big(k_M j_M[k] - \tau_0 - \kappa_\tau\big(\hat{\varpi}[k]\big)^2 + l_\varpi\big(\hat{\varpi}[k] - \varpi_M[k]\big)\Big)$$

Thrust prefilter—A low-pass filter is applied to the desired propeller thrust $F_R[k+1]$ that yields the filtered desired thrust $F_{RF}[k+1]$ and $F_{RF}[k+2]$:

$$F_{RF}[k+2] = c_{FF} F_{RF}[k+1] + (1 - c_{FF}) F_R[k+1]. \tag{7}$$

Tracking controller—This controller is tasked with tracking the filtered desired thrust F_{RF}, i.e. to ensure that the tracking error $F_E = F_{RF} - \hat{F}$ converges. For its design the desired tracking error dynamics are set to

$$F_E[k+1] = \big(c_{FC0} + c_{FC1}\hat{\varpi}[k]\big) F_E[k], \tag{8}$$

which in combination with a (forward Euler) discretization of the model of Eq. (3e) yields the corresponding control law

$$\begin{aligned} j_M[k+1] = {} & \frac{\kappa_\tau}{\kappa_F k_M}\left(\frac{T_F[k+1]}{T_S}(F_{RF}[k+2] - F_{RF}[k+1]) + F_{RF}[k+1]\right) \\ & + \frac{\tau_0}{k_M} + K_{FC}[k+1] F_E[k+1] \end{aligned} \tag{9}$$

with $T_F[k] = \frac{\Theta_P}{2\kappa_\tau \hat{\varpi}[k]}$ and $K_{FC}[k] = \frac{\kappa_\tau}{\kappa_F k_M}\left(\frac{\Theta_P(1 - c_{FC0} - c_{FC1}\hat{\varpi}[k])}{2\kappa_\tau T_S \hat{\varpi}[k]} - 1\right)$

The choice of the closed-loop dynamics of Eq. (8) appears unconventional but is motivated by considering the following: Choosing $c_{FC1} = 0$ would lead to linear error dynamics, but would result in a high feedback gain K_{FC} at low velocities ϖ. On the other hand, choosing $c_{FC0} = 1$ would lead to a constant feedback gain K_{FC}, which is too strong at high velocities and too weak at low velocities. A practical balance between these two cases can be reached by Eq. (8).

Tuning—The concern here is with the tracking error $F_R - F$, the error of the actual thrust with respect to its reference computed from the rigid body controller. Assuming a perfect model, the estimation error $\hat{F} - F$ as well as the control error $F_{RF} - \hat{F}$ do converge independently of their input ϖ_M and F_{RF}. In this case the dynamics of the tracking error is completely determined by the prefilter of Eq. (7).

In practice, the observer gain l_ϖ is adjusted to reject the measurement noise, the control gains c_{FC0} and c_{FC1} are adjusted to counter model errors/unmodeled disturbances and the prefilter constant c_{FF} adjusts the tracking behavior. The tracking dynamics have to be "much faster" than the rigid body controller in order to justify the cascade assumption. The overall thrust control may be verified for an example measurement shown in Fig. 5 that exhibits very little control error except in the case of motor current saturation.

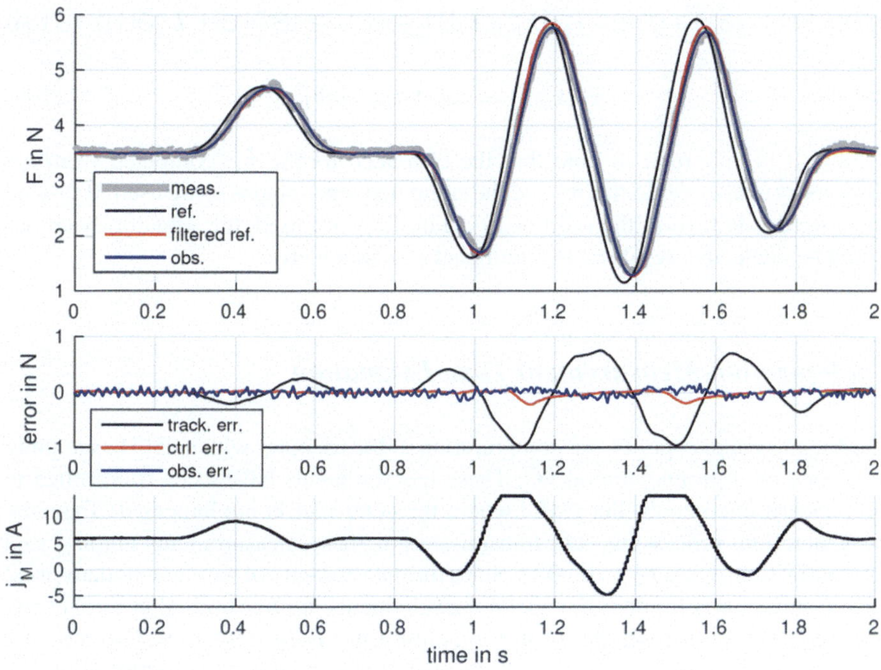

Fig. 5 Example measurement for the thrust control

State Estimation

Two kinds of measurements are used for the tricopter: An on-board inertial measurement unit (IMU) measures the vehicle's angular velocity ω_{IMU} and acceleration a_{IMU}. An external camera-based motion capture system (Vicon) measures the position r_{VIC} and orientation represented by a unit quaternion q_{VIC}. These measurements are fused as discussed in the following to estimate the quantities that are required for the rigid body controller Eq. (5a).

Angular Velocity and Torque Bias Estimation

The angular velocity measurement from the onboard gyroscopes is rather noisy and should not be used directly for feedback. While a gyro bias could also be considered, the IMU in use was found to compensate this effect internally.

The implemented observer for the angular velocity $\hat{\omega}$ and the torque bias $\hat{\tau}_B$ is a copy of the (forward Euler) discretization of the model of Eq. (3b) supplemented with a linear feedback of the velocity error $\tilde{\omega}[k] = \omega_{IMU}[k] - \hat{\omega}[k]$:

$$\hat{\omega}[k+1] = \hat{\omega}[k] + T_S \Theta^{-1}\big(\hat{\tau}_A[k] + \hat{\tau}_B[k] - \mathrm{wed}\,\hat{\omega}[k]\Theta\hat{\omega}[k] + L_\omega \tilde{\omega}[k]\big) \quad (10a)$$

$$\hat{\tau}_B[k+1] = \hat{\tau}_B[k] + L_\tau \tilde{\omega}[k]. \quad (10b)$$

Here it is important to note that the estimated torque $\hat{\tau}_A$, computed from the observed propeller velocities $\hat{\varpi}_i$ and the simulated servo angles $\hat{\theta}_i$ are used. If instead the desired values from the rigid body controller were used, the actuator dynamics would be neglected, resulting in significant estimation errors.

Configuration Measurement Time Estimation

The Vicon system captures the configuration of the tricopter with a 250Hz sampling frequency on a ground-station PC. These measurements have to be transmitted to the onboard microcontroller (MC) where the controller is implemented. Together, the generation, processing, and transmission of these measurements amount to a substantial time delay $T_D \approx 0.037$ s. Since the data capacity of wireless modems used for transmission is limited, the Vicon measurements are transmitted at a relatively low frequency corresponding to the sampling time $T_{VIC,S} \approx 0.2$ s. Furthermore, since the process involves a non-realtime operating system on the groundstation, the sampling time $T_{VIC,S}$ and the delay T_D vary with a standard deviation greater than the sampling time $T_S = 0.005$ s of the main controller.

To compensate these effects, the clocks on the PC and MC (see Fig. 6) are used as follows. The clock difference T_{PC} is constant but unknown. Since it changes whenever the PC or the MC is restarted, it varies between experiments. The delay T_D, in contrast, is variable but is assumed to have a constant mean for all experiments. These assumptions yield the relation

$$\mathrm{mean}\big(t_{REC} - t'_{VIC}\big) = T_{PC} + \mathrm{mean}(T_D)$$

where t_{REC} is the MC clock when a measurement is received and t'_{VIC} is the PC clock when the measurement was taken.

Delay identification—The mean value mean(T_D) of the time delay is estimated in the following dedicated experiment. The vehicle is mounted on an incremental

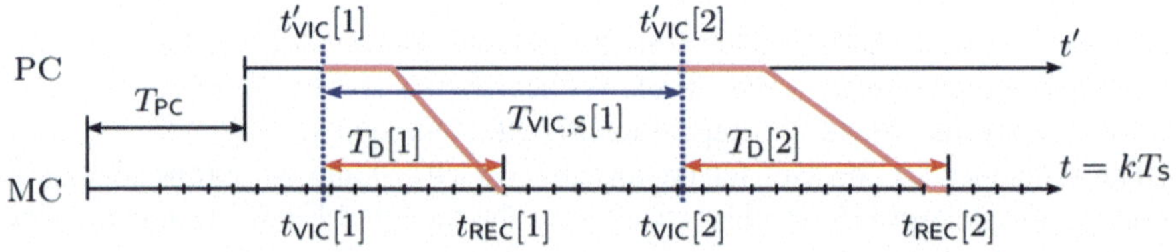

Fig. 6 Measurement time at the PC and when it is available on the microcontroller (MC)

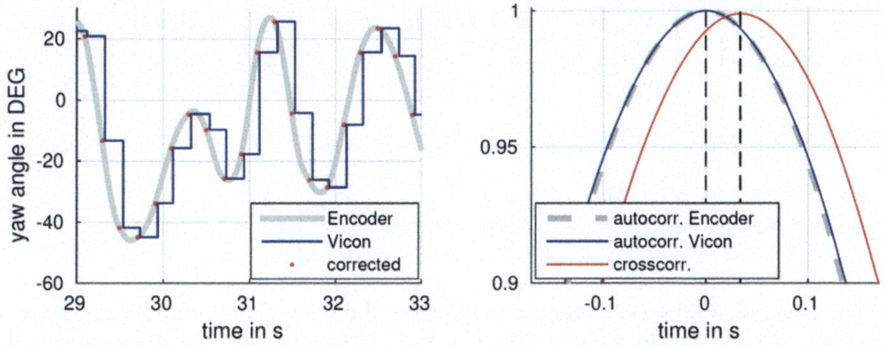

Fig. 7 Subset of encoder and Vicon data with cross-correlation for identification of time delay

encoder that serves as a reference for the *yaw*-angle. This signal and the *yaw*-angle from the motion capture system are recorded onboard and their cross-correlation is computed. A subset of this data is shown in Fig. 7 along with the relevant part of the resulting cross-correlation with maximum at mean $(T_D) = 0.034$ s.

Clock difference estimation—For real-time estimation of the clock offset $\hat{T}_{PC}$, the simple, very slow ($c_{PC} = 0.99$) low-pass filter is used:

$$\hat{T}_{PC}[k'+1] = c_{PC}\hat{T}_{PC}[k'] + (1 - c_{PC})\big(t_{REC}[k'] - t'_{VIC}[k'] - \text{mean}(T_D)\big) \quad (11)$$

Finally, the estimated time of the configuration measurement in terms of the MC clock is

$$t_{VIC}[k'] = t'_{VIC}[k'] + \hat{T}_{PC}[k']. \quad (12)$$

Configuration Estimation

With the actual time of the last configuration measurement known, the required configuration at the present sampling time may be computed using the estimated angular velocity $\hat{\omega}$ and accelerometer measurement a_{IMU}. These are related to the configuration, parameterized by the position $r \in \mathbb{R}^3$, and the unit quaternion $q = [q_w, q_x, q_y, q_z]^T \in \mathbb{R}^4$, $q^T q = 1$ by

$$\dot{q} = A_{quat}(q)\hat{\omega}, \ q\big(t_{VIC}[k']\big) = q_{VIC}[k'], \quad (13a)$$

$$\ddot{r} = R_{quat}(q)(a_{IMU} - a_B) + a_G, \ r\big(t_{VIC}[k']\big) = r_{VIC}[k'], \quad (13b)$$

with a slowly varying accelerometer bias a_B and

$$R_{quat}(q) = \begin{bmatrix} 1-2(q_y^2+q_z^2) & 2(q_yq_x - q_wq_z) & 2(q_zq_x + q_wq_y) \\ 2(q_yq_x + q_wq_z) & 1-2(q_x^2+q_z^2) & 2(q_yq_z - q_wq_x) \\ 2(q_zq_x - q_wq_y) & 2(q_yq_z + q_wq_x) & 1-2(q_x^2+q_y^2) \end{bmatrix},$$

$$A_{quat}(q) = \begin{bmatrix} -q_x & -q_y & -q_z \\ q_w & -q_z & q_y \\ q_z & q_w & q_y \\ -q_y & q_x & q_w \end{bmatrix}.$$

The basic idea now is to integrate (13) from the known configuration measurement at t_{VIC} until the timestep at $t = (k+1)T_S$. This integration is straightforward for the attitude dynamics of Eq. (13a). For the position dynamics of Eq. (13b), however, initial conditions for the velocity $\dot{r}$ and accelerometer bias a_B are required. These are provided by an observer that is sketched in Fig. 8. A core ingredient of

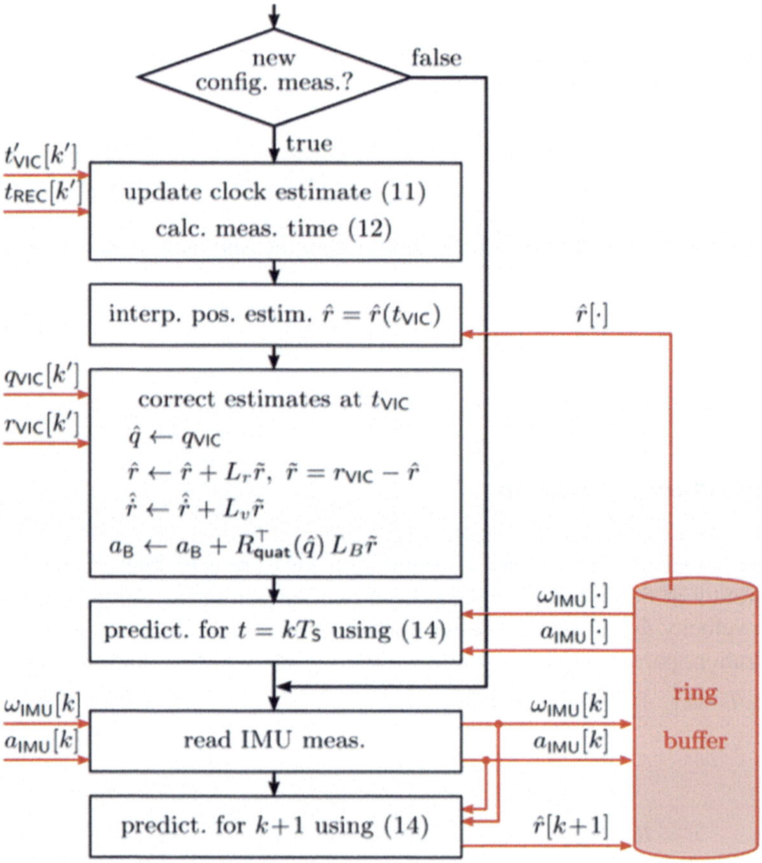

Fig. 8 Flow chart of the configuration estimation algorithm

the implementation is a ring buffer that stores the last 64 samples of the angular velocity $\hat{\omega}$, the acceleration a_{IMU}, and the estimated position $\hat{r}$. When a new configuration measurement (r_{VIC}, q_{VIC}) is available, the following is executed: First the corresponding time t_{VIC} in terms of the MC clock is estimated. Then the previously estimated position $\hat{r}(t_{VIC})$ is interpolated from the buffered values. The error w.r.t. the measurement $\tilde{r} = r_{VIC} - \hat{r}$ is used to correct the estimates $\hat{r}(t_{VIC})$, $\hat{\dot{r}}(t_{VIC})$ and $\hat{a}_B(t_{VIC})$ at the measurement time t_{VIC}. These are the initial conditions for numerically integrating the current configuration using the (forward Euler) discretization of (13):

$$\hat{q}[k+1] = \hat{q}[k] + T_S A_{quat}(\hat{q}[k])\hat{\omega}[k] - \lambda_{quat}\Big((\hat{q}[k])^T\hat{q}[k] - 1\Big)\hat{q}[k] \quad (14a)$$

$$\hat{r}[k+1] = \hat{r}[k] + T_S\hat{\dot{r}}[k], \quad (14b)$$

$$\hat{\dot{r}}[k+1] = \hat{\dot{r}}[k] + T_S\big(R_{quat}(\hat{q}[k])\big(a_{IMU}[k] - \hat{a}_B[k]\big) + a_G\big). \quad (14c)$$

The last part of (14a) is a stabilization (feedback gain $\lambda_{quat} > 0$) of the condition $\hat{q}^T\hat{q} = 1$ to counter numerical integration error. For most sampling steps, no new configuration measurement is available. In this case, the estimates $\big(\hat{q}[k+1], \hat{r}[k+1], \hat{\dot{r}}[k+1]\big)$ are only integrated. Since these values are corrected when a new measurement arrives after a finite time, this prediction remains bounded.

The performance of the configuration estimator is shown in Fig. 9. Note that the displayed acceleration a_x is expressed in the body-fixed frame, while the velocity $\hat{\dot{r}}$ is expressed in the inertial frame. These quantities are therefore not directly related. To note, however, is that estimates $\hat{r}_x$ and $\hat{q}_x$ are smooth despite the low sampling rate of the Vicon measurements. As a result, the estimator predictions must be accurate since corrections are small when they become available.

Force Bias Estimation

The combination of the model kinetics of Eq. (3a) and the accelerometer model Eq. (13b) is

$$m(a_{IMU} - a_B) = F_A + F_B.$$

From the accelerometer measurement a_{IMU} and the estimates for the accelerometer bias $\hat{a}_B$, the propeller force $\hat{F}_A$ from the previous subsections, the disturbance force F_B could be computed directly. Due to significant noise on the accelerometer measurement a_{IMU}, the value is low-pass filtered to yield the discrete estimator

$$\hat{F}_B[k+1] = L_F\hat{F}_B[k] + (I_3 - L_F)\Big(\hat{F}_A[k] - m\big(a_{IMU}[k] - \hat{a}_B[k]\big)\Big). \quad (15)$$

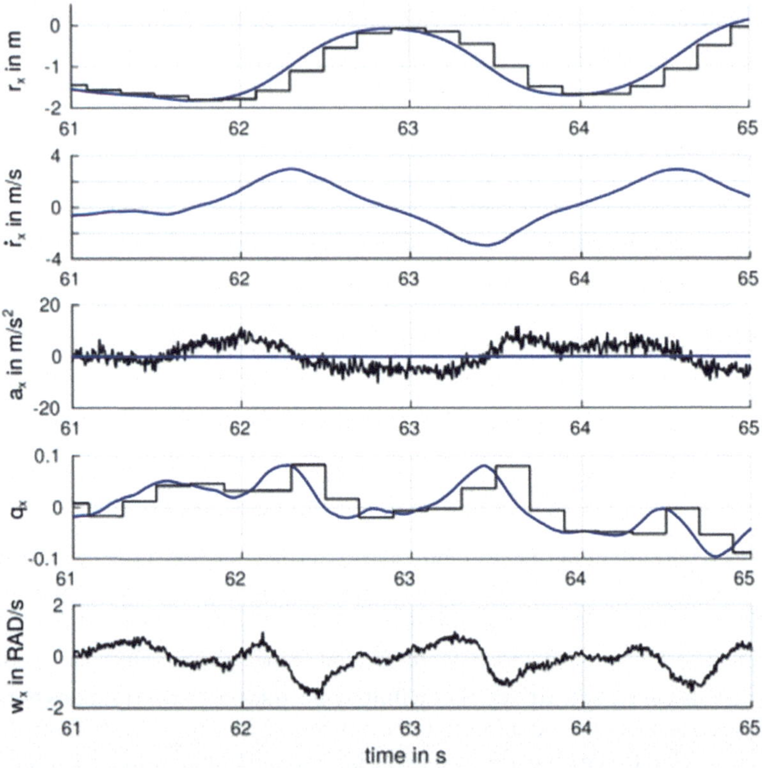

Fig. 9 Experimental estimation results (blue) and measurements (black)

Flight Test Results

In order to validate the presented algorithms, transitions in position and attitude are commanded with respective experimental results[1] shown in Fig. 10. Sufficiently smooth polynomial reference trajectories are computed on-board the tricopter based on endpoint configuration and transition time parameters sent from the groundstation.

For display convenience, the attitude is represented by Euler angles in the *roll-pitch-yaw* convention, though this parameterization is not used internally. To quantify the tracking error the position error e_r and attitude error e_R are considered according to:

$$e_r = ||r - r_R||, e_R = \cos^{-1} \frac{1}{2}\big(\mathrm{tr}\big(R_R^T R\big) - 1\big)$$

For the first position transitions starting around $t = 2$ s in Fig. 10, the constant reference attitude $R_R = I_3$ is set. For the second transitions after $t = 7$ s, the reference

[1]See also https://youtu.be/oS5PHr6H0K4.

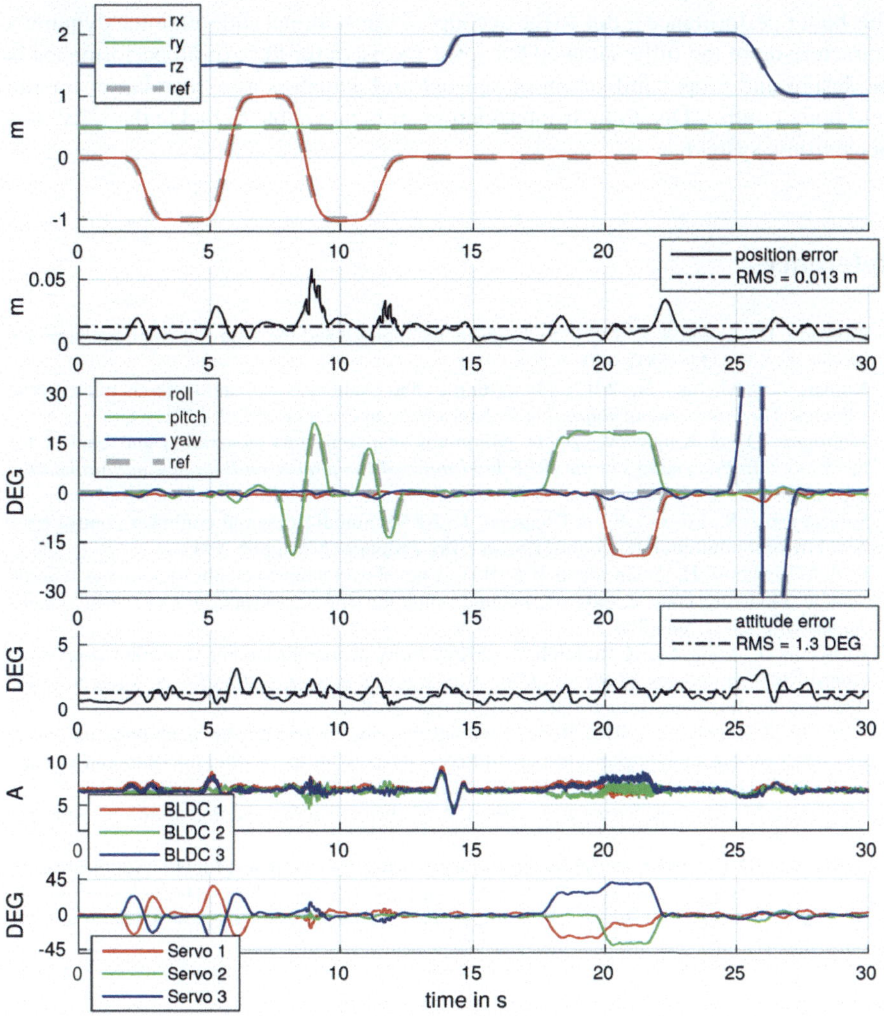

Fig. 10 Experimental position and attitude transitions tracking results

attitude R_R is set such that the resulting body-fixed force in the x and y directions vanish. This choice results in a transition characteristic of conventional multi-copters without tiltable propellers (The servo angles still vary slightly since they are used by the controller to counter disturbances). The second part of Fig. 10, transitions of the attitude are shown while the position is maintained. The last transition involving a full rotation (360° in 3 s) about the vehicle's vertical axis demonstrates the high *yaw* rates achievable with the tricopter.

The overall the tracking behaviour of the tricopter, in particular the position accuracy, is better than the one of a comparable quadcopter also developed at the LSR.

This better performance is due to the decoupled translational and rotational dynamics characteristic of the fully-actuated tricopter. Decisive for the overall performance is the design and proper utilization of sensors and actuators as a foundation for the rigid body control. Due to their interrelations it is inevitable to tackle the UAV as a mechatronic system.

References

1. Floreano, D., & Wood, R. J. (2015). Science, technology and the future of small autonomous drones. *Nature, 521*(7553), 460.
2. Kumar, V., & Michael, N. (2012). Opportunities and challenges with autonomous micro aerial vehicles. *The International Journal of Robotics Research, 31*(11), 1279–1291.
3. Mellinger, D., & Kumar, V. (2011). Minimum snap trajectory generation and control for quadrotors. In *Proceedings of the IEEE International Conference on Robotics and Automation* (pp. 2520–2525).
4. Salazar-Cruz, S., Lozano, R., & Escareño, J. (2009). Stabilization and nonlinear control for a novel trirotor mini-aircraft. *Control Engineering Practice, 17*(8), 886–894.
5. Ryll, M., Bilthoff, H., & Giordano, P. (2014). A novel overactuated quadrotor unmanned aerial vehicle: Modeling, control, and experimental validation. *IEEE Transactions on Control Systems Technology, 23*(2), 540–556.
6. Kastelan, D., Konz, M., & Rudolph, J. (2015). Fully actuated tricopter with pilot-supporting control. In *Proceedings of the 1st IFAC Workshop on Advanced Control & Navigation for Autonomous Aerospace Vehicles ACNAAV'15* (pp. 79–84).
7. Konz, M., & Rudolph, J. (2018). Redundant configuration coordinates and nonholonomic velocity coordinates in analytical mechanics. In *Proceedings of the 9th Vienna International Conference on Mathematical Modelling* (pp. 409–414).
8. Murray, R. M., Li, Z., & Sastry, S. S. (1994). *A Mathematical Introduction to Robotic Manipulation*. CRC Press.
9. Koditschek, D. E. (1989). The application of total energy as a Lyapunov function for mechanical control systems. In J. E. Marsden, P. S. Krishnaprasad, & J. C. Simo (Eds.), *Dynamics and control of multibody systems, contemporary mathematics, American Mathematical Society* (Vol. 97, pp. 131–157).

Simplified Manufacturing of Machine Tools Utilising Mechatronic Solutions on the Example of the Experimental Machine MAX

Steffen Ihlenfeldt, Jens Müller, Marcel Merx, Matthias Kraft and Christoph Peukert

Abstract This chapter presents a mechatronic system concept for highly productive and accurate machine tools. Using the example of an experimental machine called 'MAX', it will be demonstrated that the working precision of a machine can be increased whilst the effort required for machining and assembly of its mechanical components is kept to a minimum. Firstly, a novel machine structure is introduced, which allows a high reproducibility with low manufacturing effort and provides the necessary degrees of freedom for the correction of motion deviations as well as the additional actuators required for the compensation of dynamic excitations. The simplified requirements for production and assembly of the machine presented leads inevitably to large geometric and kinematic errors. These errors are modelled using rigid-body kinematics. Circularity tests and angular measurements are performed to verify the machine's ability to correct its geometric-kinematic deviations. For a highly dynamic engraving process, the reduction of the dynamic excitation caused by the drive reaction forces is demonstrated using the principle of impulse compensation. Finally, an approach to the correction of elastic and thermo-elastic errors and the comprehensive modelling and simulation-based analysis of the experimental machine are outlined.

Introduction

The primary focus in machine tool design is to increase the productivity of the machines whilst improving or maintaining their machining accuracy. Furthermore, in an environment of smaller batch sizes and individualised products, high flexibility as well as low investment and operating costs are required. An increase of productivity is achievable by increasing the cutting efficiency or reducing the non-productive time. In the context of high speed machining, both can be realised by higher feed dynamics, particular higher feed acceleration and jerk. This can be achieved by reducing the

S. Ihlenfeldt (✉) · J. Müller · M. Merx · M. Kraft · C. Peukert
Technische Universität Dresden, Dresden, Germany
e-mail: steffen.ihlenfeldt@tu-dresden.de

X.-T. Yan et al. (eds.), *Reinventing Mechatronics*,
https://doi.org/10.1007/978-3-030-29131-0_10

mass of the moving components and/or by increasing the driving forces and force-rise rates.

One way of increasing the drive forces is to use parallel drives, which are frequently found in machine tools, e.g. as gantry configurations. They enable higher feed dynamics and the "Drive at Center of Gravity" (DCG) according to Hiramoto et al. [1]. In addition, linear motors are often used to increase the feed dynamics of machine tools. In order to handle the structural excitation resulting from highly dynamic motions, mechatronic systems (e.g. redundant axis configurations [2] as well as force [3] and impulse compensation [4]) and advanced design approaches (e.g. impulse decoupling [5, 6] and floating principle [7]) have already been investigated. An overview of methods for reducing drive-induced vibrations in machine tools can be found in Großmann et al. [8]. Since the aforementioned machine concepts are usually not adaptable to typical machining tasks, the classical machine structures still dominate the market. In order to qualify these structures to be versatile and cost-effective, a shift from massive and highly accurate mechanical components towards modular lightweight design can be observed [9–11]. Thus, recent machines provide limited static and dynamic stiffness as well as limited thermal capacity but also have to withstand higher mechanical and thermal loads.

In order to ensure overall accuracy under these unfavourable conditions, compensation and correction functionalities need to be applied. In the broader sense, compensation means that disturbance variables are directly affected by actuators [12]. For example, cooling systems (heat flow affected by active cooling) or active damping devices (exciting forces reduced by damping forces) relate to compensation strategies. In contrast, correction measures cancel out the effect caused by the disturbance variables, where the disturbance variable may produce an effect in a different physical domain [12]. For correction functionalities, normally the actuators present in the machine are used to carry out the correction movements.

Examples are the correction of thermally induced displacements or the correction of geometric-kinematic errors at the Tool Centre Point (TCP), which are affected by the machine's axes. The model-based correction algorithms developed for the parallel kinematic machines at the Department of Machine Tools Development and Adaptive Controls at the TU Dresden (LWM) provide examples of effective correction approaches [13]. Kauschinger [14] showed that deviations of motions, which occur as a result of not exactly determinable geometric parameters of the hexapod structure, thermally caused strut axis length changes as well as elastic deformations due to displaced deadweight, inertial forces or process loads, can be corrected efficiently. The prerequisites are a high reproducibility of the deviations and a sufficiently accurate model. On the example of parallel kinematics it was also shown that model order-reduced thermal finite element models can be used for control-based corrections [15].

However, state of the art machine tools mainly rely on the compensation of thermal power losses by cooling [16] and the correction of generally small volumetric errors [17]. In most cases, the correction of thermal induced errors is limited to machine tool components such as ball screws [18]. In addition, active and passive damping systems for reduction of vibrations are increasingly being used [19].

In order to enable machine tools to be more flexible and intelligent, the LWM developed the machine concept of a highly dynamic over-actuated lightweight machine and implemented it in the experimental machine 'MAX' [20]. Some of the aforementioned correction and compensation approaches are implemented in this machine. As an exemplar, the correction of geometric-kinematic deviations and compensation of dynamic errors is presented in this chapter. In contrast to the classic design approach, this novel concept is intended to consistently shift the precision-relevant system functionalities from the machine structure to the mechatronic subsystem. Thus, the lightweight structure is designed to be modular and easy to manufacture and to assemble. This 3-axis Cartesian machine enables correction functionalities by means of structure-integrated sensors in 6 Degree of Freedom (DoF) through over-actuation by means of parallel drives connected via elastic coupling elements. In order to achieve the required geometric-kinematic, elastostatic and thermo-elastic accuracy, correction algorithms can be implemented in the machine's powerful PC-based control system. Additional compensation drives facilitate active vibration reduction.

In Section "Design of the Experimental Machine MAX", the novel system concept and the mechanical design of the experimental machine MAX are presented. The correction of the geometric and kinematic errors in 6 DoF associated with the simple and inexpensive construction is addressed in Section "Correction of Geometric and Kinematic Errors". The principle of impulse compensation for the reduction of vibration excitation whose efficiency is proven in a machining experiment is presented in Section "Compensation of Dynamic Errors".

Design of the Experimental Machine MAX

Concept and Design Goal of the Experimental Machine

The experimental machine illustrated in Fig. 1, is characterised by a simple and flexibly configurable design. The components of this cost-effective solution are easy to manufacture and to assemble. The chosen design approach is based on a reduction of the moving masses. In combination with an inertia-proportional drive capacity and the application of the driving force at the most shock-insensitive position of the moving structure (centre of gravity), the excitation of vibrations due to acceleration/deceleration is kept to a minimum and high feed dynamics are achieved. In the X-Y-cross slide, ironless linear direct drives are applied, which can realise very high rates of force rise.

Unavoidably, this design approach results in an increase of dynamic and thermal loads acting on the machine with simultaneously reduced static, dynamic and thermal rigidity as well as limited geometric-kinematic accuracy of the structural components. The experimental machine is designed to ensure a high repeatability of

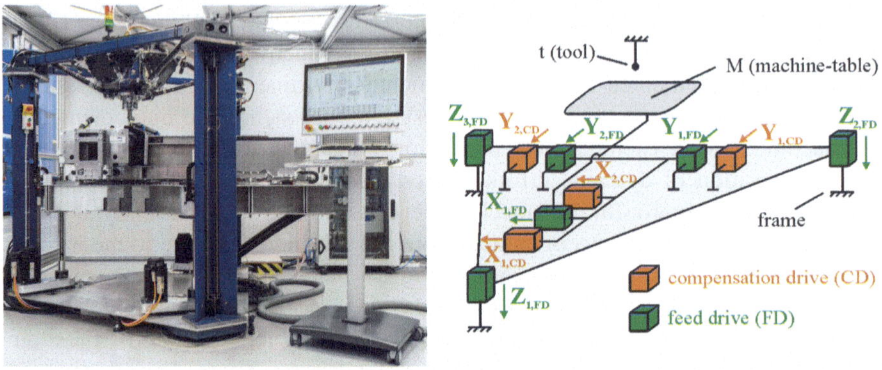

Fig. 1 Experimental machine MAX (left) and its corresponding kinematic structure (right)

the errors since this is an essential prerequisite for a successful correction of the aforementioned errors. The inaccuracies of the simple and inexpensive construction have to be eliminated by a 6 DoF correction functionality enabled by control-integrated models. In order to achieve the defined design goals, specific novel approaches and functionalities are required and all error components need to be addressed: geometric-kinematic errors, elastic errors, dynamic errors, and thermo-elastic errors.

Structure and Main Components of the Experimental Machine

The experimental machine MAX depicted in Figs. 1 and 2 realises a 3-axis Cartesian kinematic with all motions on the workpiece side. For the motion in the X- and Y-direction, ironless linear direct drives in a cross-slide arrangement are used. In the X- and Y-direction, in total four additional drives are installed to implement the impulse compensation (see Fig. 2). The reaction forces, caused by the feed forces of the slides required for machining $F_{x,FD}$, F_{y,FD_1} and F_{y,FD_2} are compensated by the counterforces F_{x,CD_1}, F_{x,CD_2}, F_{y,CD_1} and F_{y,CD_2} generated by the compensation drives. The Z-slide represents the base of the cross-slide. The masses and maximum feed forces of the slides of the experimental machine are given in Fig. 2.

The two parallel drives in Y-direction (linear motors) and the three parallel drives in Z-direction (ball screw drives) enable the execution of small correction motions in all three rotational DoF (compare with Peukert et al. [21]). Elastic coupling elements are used to unlock these DoF and to reduce geometrically and thermally induced constraining loads [22]. The compliant mechanisms are designed to allow a maximum rotation of the Z-slide around the X- and Y-axis of φ_X, $\varphi_Y = 10$ mrad. The Y-Slide can be rotated about the Z-axis by $\varphi_Z = 1.6$ mrad. The use of compliant joints permits low demands on the manufacturing accuracy of the machine assemblies and ensures high repeatability. The flexible joints of the experimental machine are discussed in more detail by Ihlenfeldt et al. [23]. The X-, Y- and Z-slide consist of water-jet cut

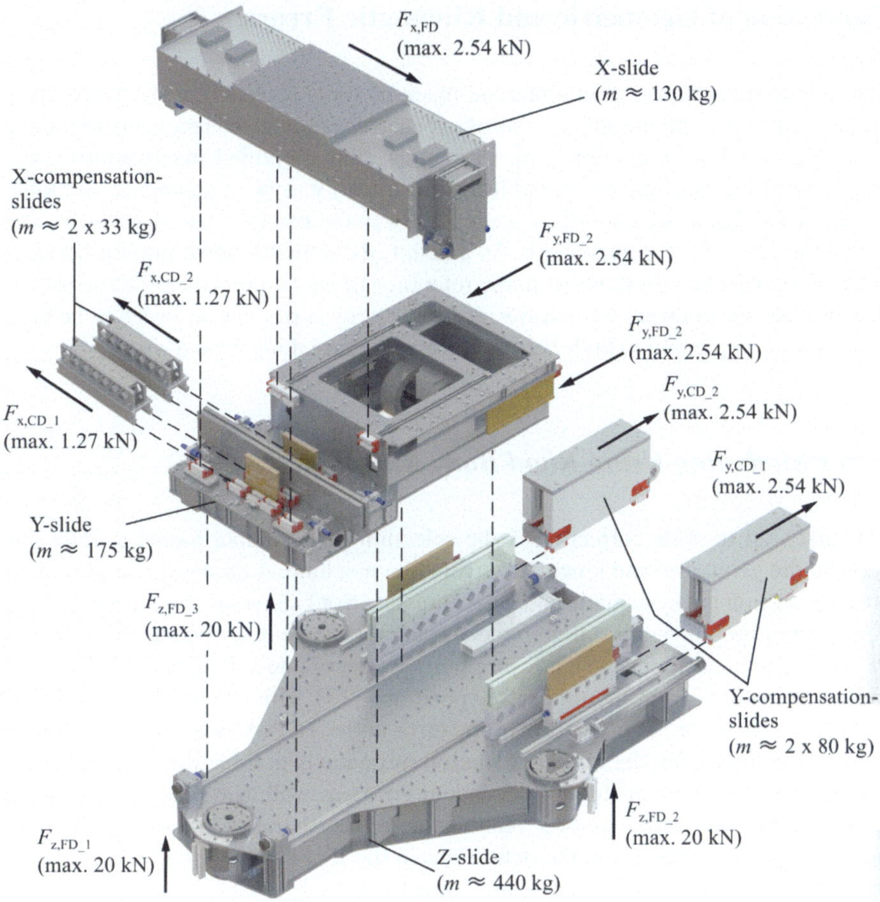

Fig. 2 Cross-slide with compensation drives in X- and Y-direction as well as the Z-slide as base

aluminium plates and ribs which are braced with tie-rods (see Peukert et al. [24]). This aluminium construction allows a simple, flexibly configurable design with high rigidity, direct load path and low mass.

The lightweight machine frame of the experimental machine is composed of three steel angle profiles and braced by tie-rods. A lightweight strut-structure holds the milling spindle as seen in Fig. 1. The machine has a large number of integrated sensors for measuring accelerations and temperatures [12]. Additionally, forces and torques are measured by using structure-integrated load cells [25]. By applying an appropriate elastic model, the force measurements can be used to correct elastic-errors of the experimental machine.

Correction of Geometric and Kinematic Errors

The geometric accuracy of commercial machine tools is usually increased by production time and with the added costs of expensive scraping and precision alignment procedures during production [26]. In contrast, the simplified manufacturing and assembly of the machine presented leads unavoidably to large geometric and kinematic errors. These are caused by inaccuracies in the machine part geometries and misalignments of the components. As a result, the actual relative motion between workpiece and tool will differ from the reference trajectory. It is therefore necessary to eliminate these errors by a software based error correction in order to achieve accuracies comparable to high performance commercial machine tools.

Error Modelling Using Rigid Body Kinematics

The first step in error correction is the selection of a suitable model to take into account the geometric and kinematic errors of a machine. A necessary condition for error correction is a systematic and deterministic error behaviour. Assuming this, the deviation of motion at a position in the workspace can be described as a function of the axes' actual positions [27]. For simplicity, errors of one axis are usually assumed to be independent from the other axes positions [17]. This allows the use of rigid body kinematics for the prediction of the error at the TCP. As a result, according to the ISO-standard 230, the motion deviation of one axis in three-dimensional space can be modelled by 6 position independent errors as well as 6 position dependent errors, each group consisting of 3 rotational and 3 translational errors [28]. The error notation used corresponds to the definitions in the ISO-standard (e.g. EYX denotes the straightness error of the X-axis in Y-direction).

According to the aforementioned definition, a 3-axis Cartesian machine tool like the experimental machine has 36 error components. By choice of a special reference system, the number of error components necessary to fully describe the kinematic behaviour can be reduced to a minimum of 21 errors: 18 position dependent errors and 3 position independent squareness errors [28].

A commonly used mathematical modelling tool for the description of rigid body movements are homogeneous transformation matrices [29]. A homogeneous transformation matrix used for kinematics modelling consists of a 3 × 3 rotation matrix ${}^{B}\boldsymbol{R}_{A}$ and a 3 × 1 translational vector ${}^{B}\boldsymbol{p}_{BA}$, embedded in a 4 × 4 matrix

$$ {}^{B}\boldsymbol{T}_{A} = \begin{pmatrix} {}^{B}\boldsymbol{R}_{A} & {}^{B}\boldsymbol{p}_{BA} \\ 0_{1\times 3} & 1 \end{pmatrix} \tag{1} $$

which transforms the coordinates of a point given in reference system A to reference system B. A consecutive transformation along a serial kinematic chain is done by multiplication of homogeneous matrices. The error at the TCP t for a X-Y–Z kinematic chain, as present in the experimental machine, can therefore be calculated from successive transformations from the machine origin M as

$$^{t^*}T_{\tilde{t}} = \left({}^{M}T_{X^*}\,{}^{X^*}T_{Y^*}\,{}^{Y^*}T_{Z^*}\,{}^{Z^*}T_{t^*}\right)^{-1}\,{}^{M}T_{\tilde{X}}\,{}^{\tilde{X}}T_{\tilde{Y}}\,{}^{\tilde{Y}}T_{\tilde{Z}}\,{}^{\tilde{Z}}T_{\tilde{t}}, \tag{2}$$

where * denotes the desired and ~ denotes the actual axis reference systems. The resulting error transformation $^{t^*}T_{\tilde{t}}$ contains rotational and translational errors.

In order to apply this serial kinematic model to the experimental machine, the actual kinematic structure, consisting of parallel drives in the Y- and Z-axes (see Fig. 1, right) is replaced by a modified kinematic structure as illustrated in Fig. 3, left. Virtual rotational axes are included in the kinematic chain, to take into account the ability to implement small changes in orientation by commanding unequal reference values to the parallel drives. The ability is realised mechanically by the use of flexible joints. These compliant mechanisms decouple the Z-slide from the guiding system and the ball-screws and are designed to allow very small rotational movements about the X- and Y-axis [23]. In order to visualise the rotational error-correction capability, the simulated elastic deformation of the flexible joints of the Z-slide due to a small rotation about the Y-axis is illustrated in Fig. 3, right.

The error transformation at the TCP t is calculated by introducing additional error transformations in the kinematic chain. For error correction, a so called decoupling method is used [30]: firstly, the rotational errors are eliminated by manipulating the commands for the parallel drives, taking into account the geometry of the experimental machine (e.g. the location of measurement systems and the pivots). In a second step, the resulting translational error vector is calculated using the extended kinematic model.

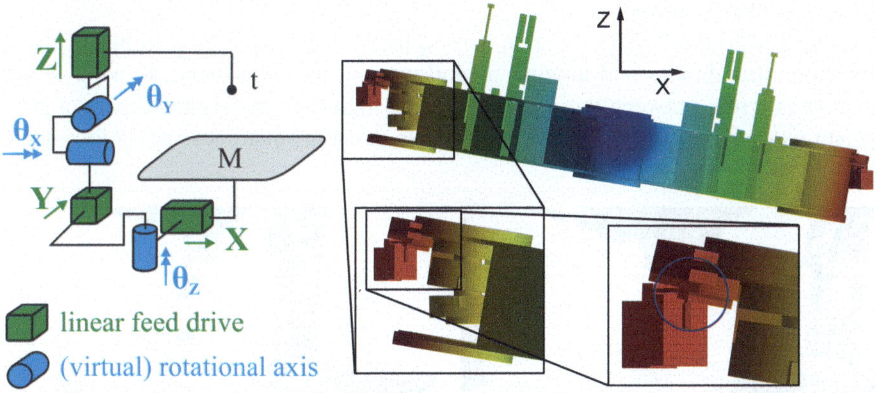

Fig. 3 Modified serial kinematic chain of the experimental machine with virtual rotational axes (left) and simulated elastic deformation of flexible joints of the Z-slide due to small rotation about the Y-axis (right)

Identification of the Geometric-Kinematic Errors

The geometric-kinematic error components were directly measured using a laser measurement system (Laser XD by Automated Precision Inc.), which utilises laser interferometry for measurement of positioning deviations and position sensitive detectors for straightness deviations [31]. Additionally, it is possible to capture angular deviations perpendicular to the laser beam optically and angular errors about the beam axis with a built-in precision inclinometer. In the case of the experimental machine, the spindle and the Z-axis are parallel. Therefore, the roll error of the Z-axis has no effect on the total volumetric error and the measurement of 20 geometric-kinematic errors is sufficient for this machine. As an exemplar, Fig. 4, left illustrates the measurement setup for the X-axis and Fig. 4, right for the squareness deviation between the X- and Z-axis.

The errors were measured using a bidirectional positioning pattern in the position range of −240 to 240 mm in the X-axis, −195 to 195 mm in the Y-axis and 0 to 240 mm in the Z-axis. The measurement results in Fig. 5 show that a linear trend is inherent to the positioning deviations of all machine axes. The proportionality factors are 272 ppm in X-, −291 ppm in Y- and 181 ppm in Z- direction. Large straightness deviations of the X-axis in Y-direction EYX with a maximum of 104 μm and straightness deviations of the Y-Axis in X-direction EXY with a maximum of 54 μm, also contribute to the inaccuracy of the machine. Due to the inaccurate assembly of the machine, the X-axis also shows large angular deviations with a maximum interval of 442 μrad about the Z-axis and 249 μrad in roll direction (see Table 1). This twisting motion is presumably a result of a parallelism errors of the X-axis guiderails [32]. The pitch error of the Y-axis EAY results probably from a change in the elastostatic deformation of the Z-slide caused by the movement of the Y-slide's centre of gravity.

Compared to the X-Y-cross-slide's axes, the Z-axis shows only minor position dependent straightness and angular deviations. On the other hand, looking at the squareness errors between the nominal axes, which were determined using a pentaprism as optical squareness artefact, reveals that the squareness deviation of the

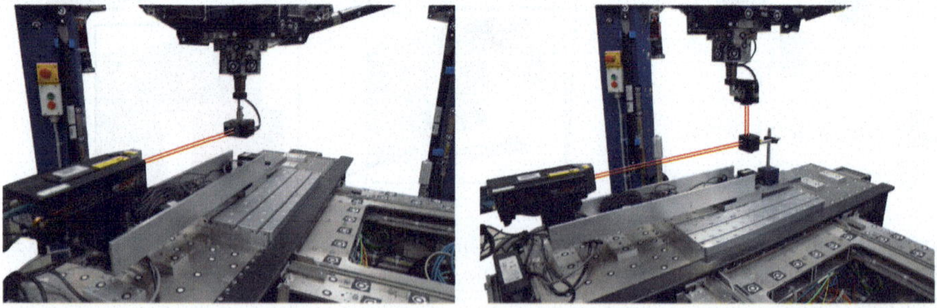

Fig. 4 Laser interferometer setup for identification of the X-axis errors (left) and the squareness deviation between the X- and Z-axis (right)

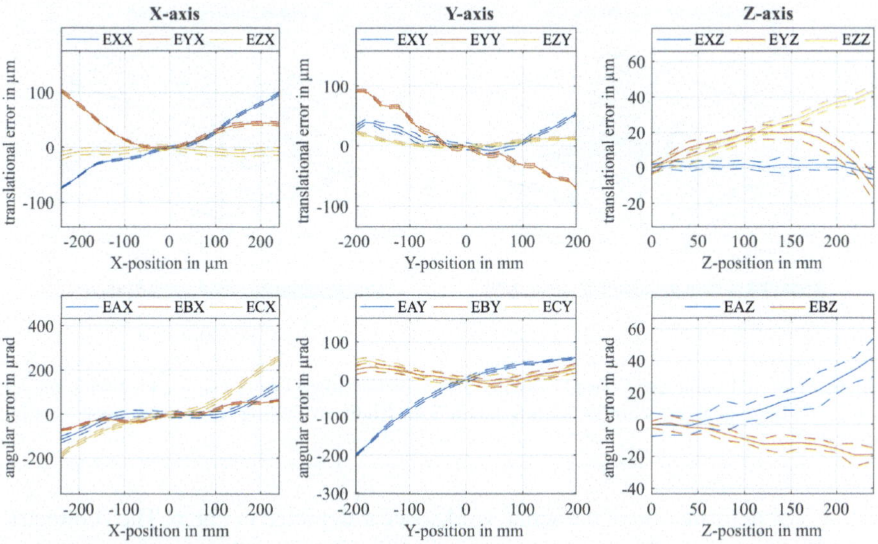

Fig. 5 Measured position dependent error components of the X-axis (left), Y-axis (centre), Z-axis (right). Solid lines: bi-directional mean. Dashed lines: 3σ confidence interval. Errors defined equal to zero at axis position 0 mm

Table 1 Maximum error interval $\Delta E = E_{max} - E_{min}$ and maximum standard deviation σ_{max} of the error component measurement data depicted in Fig. 5

X-Axis	EXX (µm)	EYX (µm)	EZX (µm)	EAX (µrad)	EBX (µrad)	ECX (µrad)
ΔE	172	105	15	249	136	442
σ_{max}	1.09	1.37	2.47	5.16	2.37	4.54
Y-Axis	EXY (µm)	EYY (µm)	EZY (µm)	EAY (µrad)	EBY (µrad)	ECY (µrad)
ΔE	61	162	24	258	44	55
σ_{max}	2.64	1.33	1.15	2.46	4.23	5.18
Z-Axis	EXZ (µm)	EYZ (µm)	EZZ (µm)	EAZ (µrad)	EBZ (µrad)	ECZ (µrad)
ΔE	6	32	31	49	28	–
σ_{max}	1.71	0.83	0.79	2.12	3.28	–

Z-axis dominates the volumetric error. While the squareness deviation between the X- and Y-axis EC0Y is relatively moderate with 176 µrad, the orientation errors of the Z-axis are large with EA0Z = 2821 µrad and EB0Z = 845 µrad. Comparing the standard deviations and the maximum error intervals, it can be stated that the machine shows a highly repeatable geometric-kinematic behaviour. Thus, the necessary condition for the correction of the observed large geometric-kinematic errors is satisfied.

The resulting volumetric error of the experimental machine was calculated as the Euclidean norm of the error vector from t^* to $\tilde{t}$ using the kinematic model described

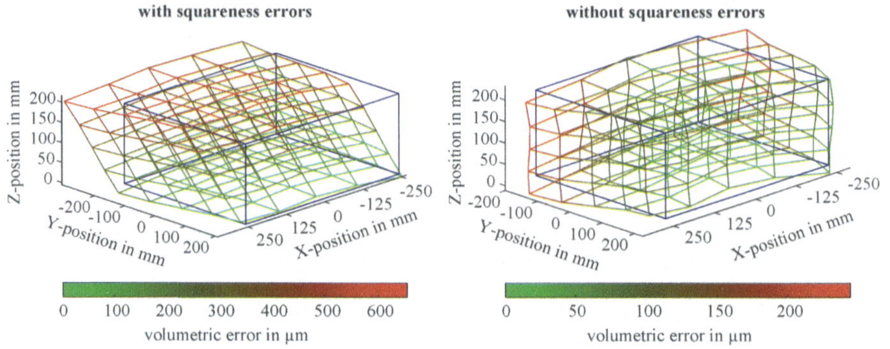

Fig. 6 Predicted volumetric error calculated from the identified errors using a kinematic model. Left: including squareness errors, scaling factor 250. Right: excluding squareness errors, scaling factor 500

earlier. The deformation of the actual workspace is depicted in Fig. 6. The volumetric error, dominated by the squareness errors of the Z-axis and the angular errors of the X-and Y-axis, has a maximum value of 655 μm, which occurs at the positive boundary of the Z-axis (see Fig. 6, left). If the squareness errors are not considered in the calculations (see Fig. 6, right), the effect of the angular errors becomes visible, since the largest volumetric errors occur at both ends of the X-axis at a Y-position of −200 mm.

Verification of the Geometric-Kinematic Error Correction

Verification of the Correction of Translational Errors

The derived real time error correction was implemented on the machine controller and parameterised with the measured error components. In order to verify the algorithm, circularity tests with and without activated geometric-kinematic error correction were conducted to evaluate the translational correction performance using a Double-Ball-Bar (DBB). In each plane, tests were performed using radii of 100, 150 and 200 mm. A low feed rate of 250 mm/min was chosen to minimise the influence of the dynamic machine behaviour. Due to geometric restrictions, half circles were commanded in the X-Z- and Y-Z-plane. The circular paths centres were set to the origin of the machine. The tool length of 226 mm differed from the one used in the laser measurements (143 mm).

During the experiments, the temperature was not constant. However, changes in the temperature only resulted in a proportional mean radius deviation, due to the thermo-symmetric design and similar expansion coefficients of the axes. Therefore, Fig. 7 illustrates the radius deviations from the least-squares circle calculated from measurement data rather than from the commanded circle. The uncorrected

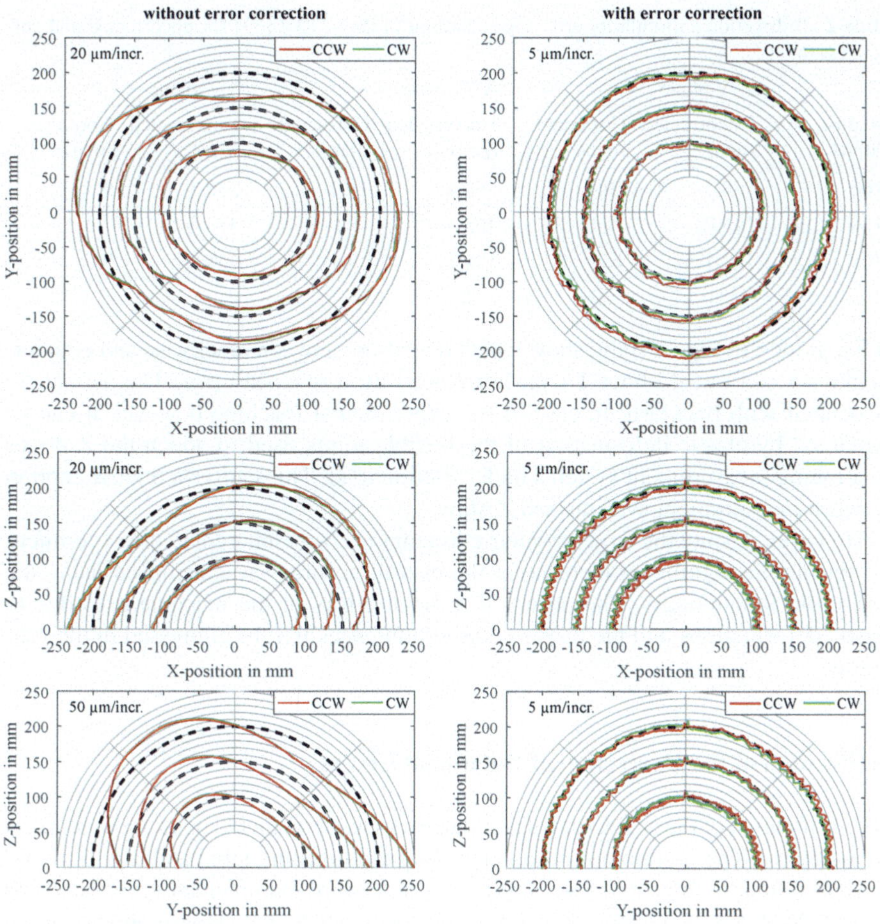

Fig. 7 Circularity test results for radii 100, 150 and 200 mm at a feed rate of 250 mm/min in the X-Y-, Y-Z- and X-Z-plane in clockwise (CW) and counter-clockwise (CCW) direction, without (left) and with error-correction (right). The centres of the circular paths were located at machine position (0, 0, 0). The tool length used was 226 mm

circularity test results are depicted in Fig. 7, left, as a reference for the geometric machine performance. The increase of circularity errors with the circle radius visible in Table 2, results mainly from the proportional positioning errors in the X-Y-plane and the squareness errors in the X-Z and Y-Z-planes. The maximum circularity errors in the uncorrected case are 151 µm in the X-Y-plane, 145 µm in the X-Z-plane and 444 µm in the Y-Z-plane and occur at a radius of 200 mm. When the error-correction is activated, the bi-directional circularity errors are reduced by 91.7% in the X-Y-plane, 93.7% in the X-Z-plane and 97.9% in the Y-Z-plane. The results of the corrected circularity tests are depicted with a much higher resolution

Table 2 Bidirectional circularity error G_b according to ISO-230-4 [33] for uncorrected and corrected paths

G_b	X-Y-plane		X-Z-plane		Y-Z-plane	
Radius (mm)	Uncorrected (μm)	Corrected (μm)	Uncorrected (μm)	Corrected (μm)	Uncorrected (μm)	Corrected (μm)
100	58.3	9.4	68.4	10.2	230.5	9.3
150	103.0	10.6	103.4	9.0	333.5	7.4
200	151.1	12.5	144.5	9.1	444.2	9.0

of 5 μm/incr. in Fig. 7, right, show a path deviation between clockwise and counter-clockwise motion of about 1 μm, introduced by a reversal error. This is usually associated with backlash. In case of the experimental machine however, it can be explained by elastic deformation of the flexible joints used to attach the Z-drives positioning measurement systems under friction load. This causes a relative motion between measurement systems and Z-slide.

However, despite the large geometric-kinematic errors inherent to the experimental machine, a significant increase in the overall accuracy was achieved using the correction functionality. The observed reversal errors of the machine axes due to elastic deformations and the thermo-elastic influences will be addressed in the near future.

Verification of the Correction of Rotational Errors

The verification of the correction of angular errors was conducted by using a precision inclinometer. The inclinometer was placed on the machine table and a grid of 100 discrete points in the calibrated working volume were commanded to the machine (four X-Y-planes in different Z-heights, 25 points per plane). Measurements of the angles about the X-axis $\varphi_X(x, y, z)$ and Y-axis $\varphi_Y(x, y, z)$ were carried out. The machine origin was defined as reference point such that $\varphi_X(0, 0, 0) = \varphi_Y(0, 0, 0) = 0$ μrad. The absolute value of the angular deviation,

$$|\varphi_{XY}| = \sqrt{\varphi_X^2 + \varphi_Y^2}, \tag{3}$$

is depicted in Fig. 8, as an exemplar, for two Z-heights in the X-Y-plane with and without geometric error correction. An increase of the angular accuracy was achieved and the maximum absolute error deviation in the considered volume was reduced by 83.2% from 316 to 53 μrad. The RMS of the angular deviation of all measurements was improved by 82.9% from 140 to 24 μrad. The remaining residual errors after correction can be explained by un-modelled interdependencies between the axes, thermal influences and measurement uncertainties of the calibration measurement.

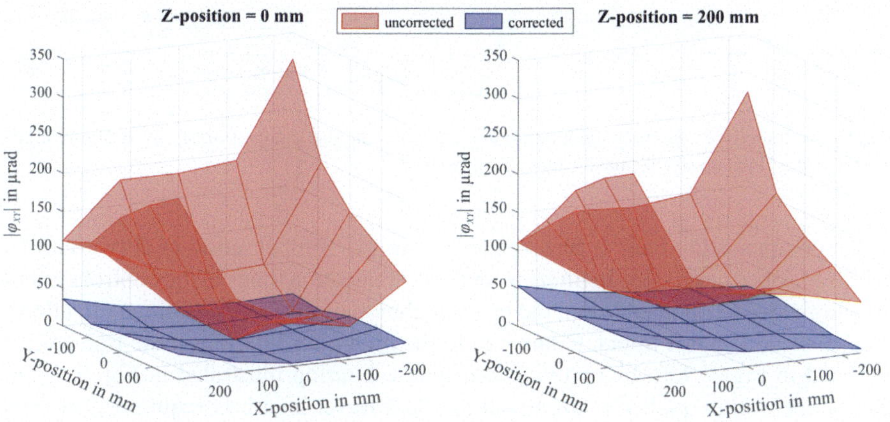

Fig. 8 Absolute angular deviation $|\varphi_{XY}|$ measured with a precision inclinometer in the X-Y-plane for z = 0 mm (left) and z = 200 mm (right)

In conclusion, it was experimentally shown that it is possible to perform small rotational correction movements by utilising flexible joints in the machine's Z-axis to significantly reduce the angular errors of the experimental 3-axis machine.

Compensation of Dynamic Errors

The experimental machine uses the principle of impulse compensation in the X-Y plane to compensate the dynamic errors that are unavoidable due to the relatively compliant machine frame. The aim is to increase the maximum permissible jerk in the trajectory while at the same time minimising the structural excitation caused by the reaction forces of the linear direct drives.

Impulse Compensation for Linear Motor Driven Feed Axes

Impulse compensation is a form of force compensation in which only the high-frequency, frame-exciting force components of the reaction force of the feed drives are compensated [34]. Low-frequency force components, which occur, for example, when moving at constant velocity, are filtered out in order to keep the travel distance of the compensation drive small. In order to avoid the limitation of the compensation effectiveness described in [35, 36] due to the limited synchronisation of the drives for the impulse compensation by means of current filtering, a variant of impulse compensation was developed, in which the target path of the compensation drive p_{CD} is calculated from the trajectories of the feed drives p_{FD}. With the offline pre-calculation

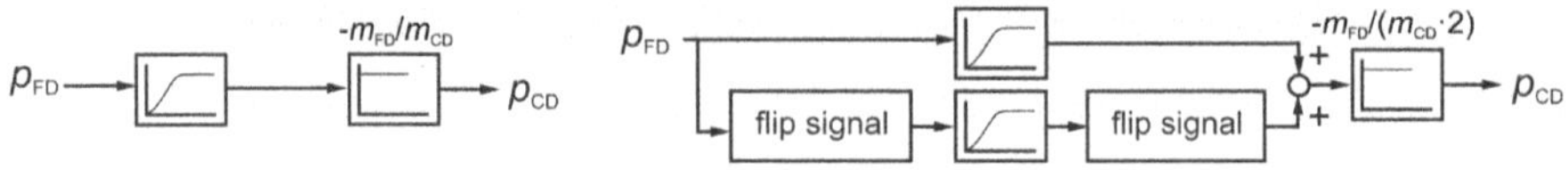

Fig. 9 Strategies for trajectory-calculation to realise the impulse compensation: by applying high-pass filtering (left) and zero-phase-filtering (right)

of the compensation path, a further increase in the compensation quality could be achieved, since delays due to filtering and communication times in the control system are eliminated [37]. In addition, the pre-calculation of the compensation path allows the use of non-causal filters. Compared to the high-pass-filtering of the trajectories depicted in Fig. 9, left, by applying zero-phase-filtering illustrated in Fig. 9, right, the compensation quality is improved and less travel length is required [38]. The trajectories of the compensation drives of the cross-slide are calculated taking into account the centre of gravity of the slides to avoid excitation of yaw-vibrations [38]. An overview of the implementation possibilities of the impulse compensation is given by Peukert, Müller and Großmann [39]. In Ihlenfeldt et al. [23], the compensation effect was demonstrated by means of a milling process. In the following section, the compensation of dynamic errors by means of an engraving test is presented.

Vibration Reduction Using the Example of an Engraving Process

An engraving test with high feed dynamics (velocity $v_{ref} = 0.1$ m/s, acceleration $a_{ref} = 5$ m/s^2 and jerk $j_{ref} = 5000$ m/s^3) was performed. The trajectories of the compensation drives were calculated by applying zero-phase-filtering and considering the mass ratios (see [23]) according to Fig. 9, right. The zero-phase-filtering was based on a second order Butterworth filter (high-pass) with a cut-off frequency of $f_{c,x} = 0.1$ Hz (X-axis) and $f_{c,y} = 0.05$ Hz (Y-axis). The engraving tool and a white painted aluminium plate, which served as the workpiece are depicted in Fig. 10. After the engraving process, blue spotting paste was applied to make the contour visible. As can be seen in Fig. 10, centre, strong vibrations are recognisable without compensation. With active impulse compensation, the vibrations can be reduced significantly as depicted in Fig. 10, right. The remaining contour error (overshoot at the corners) does not result from structural oscillations. By optimising the controller settings of the feed drives and implementing an acceleration feedforward control, the contour accuracy could be further improved.

The engraving tests pointed out that the vibrational excitation can be significantly reduced by active compensation, even within a dynamically very compliant machine structure.

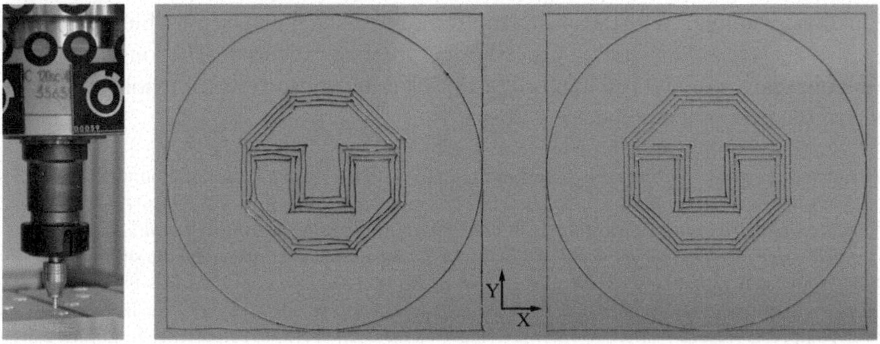

Fig. 10 Engraving test: tool (left) as well as results without (centre) and with impulse compensation (right)

Conclusion and Outlook

In this chapter, a novel design concept for machine tools for highly dynamic and accurate milling tasks was introduced. This concept is intended to consistently shift the precision-relevant, accuracy-determining machine characteristics from the mechanical to the mechatronic domain. Thus, the use of an economically manufactured lightweight machine structure with limited basic accuracy, which nevertheless offers good reproducibility of the system's inherent errors, becomes possible. In addition, the amount and arrangement of the feed drives enables the correction of geometric-kinematic and thermally induced deviations at the TCP. The dynamic structure excitation resulting from the drive reaction forces is minimised by means of compensation drives counteracting synchronously to the motions of the feed drives. The modelling and correction of the geometric-kinematic deviations by means of a rigid body kinematic approach was introduced and experimentally proven by means of circularity tests as well as angular measurements. A reduction of the original deviations of approximately 90% could be verified. The significant reduction of the drive-induced structural vibrations that can be achieved by means of impulse compensation was demonstrated by a highly dynamic engraving process.

In the near future, the experimental machine‘s control will be equipped with the algorithms for the correction of thermally induced displacements at the TCP. These algorithms are based on order-reduced thermo-elastic finite element models and thus enable a very accurate but sufficiently fast calculation of the temperature field on the machine control. From the current temperature field the deformation of the structural components and finally the error resulting at the TCP is determined in 6 degrees of freedom. Based on the kinematic model presented in this chapter, the correction values are calculated.

In addition to the thermo-elastic model [40], an elastic multi-body model of the experimental machine was created in MATLAB/Simulink. The dynamic and elastic characteristics of the components are taken into account by applying a modal order reduction technique. With the elastic multi-body model, it will be possible to perform detailed analyses of machining processes before their execution. The long-term goal is to combine the models of all physical domains and thus to create an holistic digital twin of the experimental machine 'MAX'.

Acknowledgements The presented work was funded by the German Research Foundation (DFG) within the projects GR1458/48-3 "*Basic investigations on the application of the pulse compensation for linear direct drives in a cross-slide*", GR1458/64-1 "*Application potential of articulated coupled drive and guide elements for increase of movement dynamics and accuracy*" and the subproject C06 in the Collaborative Research Centre SFB/Transregio 96 "*Thermo-energetic Design of Machine Tools*". The authors gratefully thank the DFG for their generous support. Additional thanks go to Maria Kopp, Evi Karola Wörner, Maria Meier, Jessica Deutsch, Bertram Friedrich, Mingliang Yang, Ziyi Wang, Axel Fickert, Felix Bender, Ulysse Delplanque, Jiajun Ruan, Daniel Sandoval Ovalle, Xaver Thiem, Enrico Henschel, Sven Kung, Bin Zhou, Luca Di Giorgio, Simon Städtler and Holger Kretzschmar who supported us in developing the models, designing of components and performing experiments.
The chapter is dedicated to Prof. Dr.-Ing. habil. Knut Großmann, who developed the idea for the design concept.

References

1. Hiramoto, K., Hansel, A., Ding, S., & Yamazaki, K. (2005). A Study on the drive at center of gravity (DCG) feed principle and Its application for development of high performance machine tool systems. *CIRP Annals, 54*(1), 333–336.
2. Ihlenfeldt, S., Müller, J., Peukert, C., & Merx, M. (2016). Kinematically coupled force compensation — design principle and control concept for highly-dynamic machine tools. *Procedia CIRP, 46,* 189–192.
3. Hollis, R. L., Jr. (1993). *Ultrafast Electro-Dynamic X, Y and Theta Positioning Stage*. European patent specification EP0523042B1.
4. Großmann, K., Müller, J., Jungnickel, G., & Mühl, A. (2006). *Impulskompensation*. German patent DE 10 2004 057 062 B4.
5. Brecher, C., Wenzel, C., & Klar, R. (2008). Characterization and optimization of the dynamic tool path of a highly dynamic micromilling machine *CIRP. Journal of Manufacturing Science & Technology, 1*(2), 86–91.
6. Müller, J., Junker, T., Pagel, K., Großmann, K., & Drossel, W. -G. (2013). Impulsentkopplung von Lineardirektantrieben, *wt Werkstattstechnik online, 103*(5), 370–376.
7. Bubak, A., Soucek, P., & Zelený, J. (2003). New principles for the design of highly dynamic machine tools. In *Proceedings of the 17th International Conference on Production Research ICPR-17* (pp. 1–10). Blacksburg.
8. Großmann, K., Müller, J., Merx, M., & Peukert, C. (2014). Reduktion antriebsverursachter Schwingungen. *Antriebstechnik/Ant Journal, 53*(4), 35–42.
9. Kroll, L., Blau, P., Wabner, M., Frieß, U., Eulitz, J., & Klärner, M. (2011). Lightweight components for energy-efficient machine tools. *CIRP Journal of Manufacturing Science & Technology, 4*(2), 148–160.
10. Möhring, H.-C., Brecher, C., Abele, E., Fleischer, J., & Bleicher, F. (2015). Materials in machine tool structures. *CIRP Annals, 64*(2), 725–748.
11. Zulaika, J., & Campa, F. J. (2009). New concepts for structural components. In N. Lopez de Lacalle & A. Lamkiz Mentxaka (Eds.), *Machine tools for high performance machining*. Springer.

12. Großmann, K. (Ed.) (2014). *Thermo-energetic design of machine tools*. Springer.
13. Großmann, K., Kauschinger, B., & Szatmári, S. (2008). Kinematic calibration of a hexapod of simple design. *Production Engineering Research & Development, 2,* 317–325.
14. Kauschinger, B. (2006). *Verbesserung der Bewegungsgenauigkeit an einem Hexapod einfacher Bauart*, Dissertation, Technische Universität Dresden, Germany.
15. Beitelschmidt, M., Galant, A., Großmann, K., & Kauschinger, B. (2015). Innovative simulation technology for real-time calculation of the thermo-elastic behaviour of machine tools in motion. *Applied Mechanics & Materials, 794,* 363–370.
16. Weber, J., Weber, J., Shabi, L., & Lohse, H. (2016). Energy, Power and heat flow of the cooling and fluid systems in a cutting machine tool. *Procedia CIRP, 46,* 99–102.
17. Schwenke, H., Knapp, W., Haitjema, H., Weckenmann, A., Schmitt, R., & Delbressine, F. (2008). Geometric error measurement and compensation of machines—An update. *CIRP Annals, 57(2),* 660–675.
18. Zhang, J., Li, B., Zhou, C., & Zhao, W. (2016). Positioning error prediction and compensation of ball screw feed drive system with different mounting conditions. *Proceedings of the Institution of Mechanical Engineers, Part B: Journal of Engineering Manufacture, 230(12),* 2307–2311.
19. Brecher, C., Baumler, S., & Brockmann, B. (2013). Avoiding chatter by means of active damping systems for machine tools. *Journal of Machine Engineering, 13*(3), 117–128.
20. Großmann, K., Möbius, V., Höfer, H., Müller, J., & Kauschinger, B. (2009). German patent DE 10 2009 057 207 A1.
21. Peukert, C., Merx, M., Müller, J., Kraft, M., & Ihlenfeldt, S. (2017). Parallel-driven feed axes with compliant mechanisms to increase dynamics and accuracy of motion. In R. Schmidt, G. Schuh (Eds.), *Proceedings of the 7th WGP-Jahreskongress* (pp. 417–424). Aachen: Apprimus.
22. Peukert, C., Merx, M., Müller, J., & Ihlenfeldt, S. (2017). Flexible coupling of drive and guide elements for parallel-driven feed axes to increase dynamics and accuracy of motion. *Journal of Machine Engineering, 17*(2), 77–89.
23. Ihlenfeldt, S., Müller, J., Merx, M., & Peukert, C. (2018) A novel concept for highly dynamic over-actuated lightweight machine tools. In X. Yan, D. Bradley, P. Moore (Eds.). *Reinventing Mechatronics: Proceedings of Mechatronics 2018* (pp. 210–216). Glasgow.
24. Peukert, C., Müller, J., Merx, M., Galant, A., Fickert, A., Zhou, B., Städtler, S., Ihlenfeldt, S., & Beitelschmidt, M. (2018) Efficient FE-modelling of the thermo-elastic behaviour of a machine tool slide in lightweight design. In *Proceedings of the 1st Conference on Thermal Issues in Machine Tools*. Dresden.
25. Friedrich, C., Kauschinger, B., & Ihlenfeldt, S. (2016). Decentralized structure-integrated spatial force measurement in machine tools. *Mechatronics, 40,* 17–27.
26. Sartori, S., & Zhang, G. X. (1995). Geometric error measurement and compensation of machines. *CIRP Annals, 44*(2), 599–609.
27. Weck, M., & Brecher, C. (2006). *Werkzeugmaschinen 5—Messtechnische Untersuchung und Beurteilung, dynamische Stabilität* (7th edn., pp. 80–87). Springer.
28. International Standards Organization. (2012). *ISO 230-1:2012(E), Test code for machine tools—Part 1: Geometric accuracy of machines operating under no-load or quasi-static conditions* @ www.iso.org/obp/ui/#iso:std:iso:230:-1:ed-3:v1:en. Accessed May 21, 2019.
29. Siciliano, B., & Khatib, O. (2016). *Handbook of robotics* (2nd edn, pp. 16–17). Springer.
30. Turek, P., Jedrzejewski, J., & Mordzycki, W. (2010). Methods of machine tool error compensation. *Journal of Machine Engineering, 10*(4), 5–25.
31. Automated Precision Inc. (2010). *User Manual XD Laser.*
32. Majda, P. (2012). Modelling of geometric errors of linear guideway and their influence on joint kinematic error in machine tools. *Precision Engineering, 36(3),* 369–378.
33. International Standards Organization. (2005). *ISO 230-4:2005(E), Test code for machine tools—Part 4: Circular tests for numerically controlled machine tools* @ www.iso.org/obp/ui/#iso:std:iso:230:-4:ed-2:v1:en. Access May 21, 2019.
34. Großmann, K., & Müller, J. (2005). Verringerung der Gestellanregung durch Lineardirektantriebe mittels Impulskompensation. *ZWF Zeitschrift für wirtschaftlichen Fabrikbetrieb, 100*(11), 656–660.

35. Großmann, K., & Müller, J. (2009). Untersuchungsergebnisse zur Wirksamkeit der Impulskompensation von Lineardirektantrieben. *ZWF Zeitschrift für wirtschaftlichen Fabrikbetrieb, 104*(9), 761–767.
36. Müller, J. (2009). *Vergleichende Untersuchung von Methoden zur Verringerung der Gestellanregung durch linearmotorgetriebene Werkzeugmaschinenachsen*, Dissertation, Technische Universität Dresden, Germany.
37. Großmann, K., Müller, J., & Peukert, C. (2011). Vorab-Sollwertberechnung für die Impulskompensation von Lineardirektantrieben. *ZWF Zeitschrift für wirtschaftlichen Fabrikbetrieb, 106*(5), 352–355.
38. Peukert, C., Müller, J., Merx, M., Kung, S., & Großmann, K. (2015). Sollbahnberechnung zur Impulskompensation eines linearmotorgetriebenen Kreuzschlittens. In *VDI-Berichte 2268: Antriebssysteme 2015* (pp. 117–132). Düsseldorf: VDI.
39. Peukert, C., Müller, J., & Großmann, K. (2012). Sollbahnberechnung zur Impulskompensation von Lineardirektantrieben. In *VDI-Berichte 2175: Bewegungstechnik 2012* (pp. 171–184). Düsseldorf: VDI.
40. Thiem, X., Kauschinger, B., & Ihlenfeldt, S. (2018). Structure model based correction of machine tools. In *Proceedings of the 1st Conference on Thermal Issues in Machine Tools*. Dresden.

Simulation in the Design of Machine Tools

Daniel Spescha, Sascha Weikert and Konrad Wegener

Abstract A framework of methods for efficient and accurate simulation of the dynamics of machine tools including control is presented. The major achievements are a model order reduction technique with pre-definable error bound and a method for modelling of moving interfaces on flexible bodies based on trigonometric interpolation of the desired force distribution. The software tool MORe (**M**odel **O**rder **R**eduction and mor**e**) that implements these methods is presented. Application examples on analyses of dynamic and static properties are presented and simulation results are compared with measurements. The very good validation results confirm the usability of the presented methods and software for real-world applications.

Introduction

Machine tools, as a special class of mechatronic systems, have been successfully designed and optimised using physical prototypes and empirical methods for decades. In order to reinvent machine tools, new concepts in terms of both mechanics and control have to be tested and evaluated during the design process. This requires efficient methods, algorithms, and software for the simulation of dynamic and static properties of machine tools. In the following, methods and a software solution for efficient simulation of the dynamics of machine tools using accurate reduced order models is presented.

D. Spescha (✉) · S. Weikert
Inspire AG, Zurich, Switzerland
e-mail: spescha@inspire.ethz.ch

K. Wegener
ETH Zürich, Zurich, Switzerland

X.-T. Yan et al. (eds.), *Reinventing Mechatronics*,
https://doi.org/10.1007/978-3-030-29131-0_11

State of the Art in Machine Tool Simulation

In 2005, Altintas et al. [1] illustrated the benefit of virtual machine tools in order to reduce the development time and thus enhance the competitive ability of machine tool manufacturers. They stated that time- and cost-intensive manufacturing, testing, and optimisation of physical prototypes can no longer be afforded and can, at least partly, be replaced by virtual prototyping and verification. Applications for virtual prototypes reach from simulation of feed drives as pointed out, for instance by Altintas et al. [2] to simulation of the interaction of the manufacturing process and machine tool as shown by Brecher et al. [3].

Nevertheless, the use of virtual prototypes has not yet become extensively popular in the machine tool industry. Even though model-based calculation of specific properties of single machine components such as static stiffness, maximum stress, or eigenfrequencies, is commonly used, comprehensive models of whole machine tool structures are rarely used.

In academic environments, some attempts aimed at the creation of virtual prototypes have been published. Weikert [4] presented an application oriented simulation of machine tools and its components, whereas the structure was modelled as an assembly of rigid bodies. Berkemer [5] investigated the coupled simulation of structural mechanics, drives, and control loop of machine tools using flexible multi-body simulation. For model reduction, component mode synthesis methods were proposed. Zaeh and Siedl [6] investigated simulation methods for the dynamic behaviour of machine tools during axis movements. For the first time, the problem of moving interfaces and effects of oscillation due to finite element discretisation was addressed. A combination of reduced models derived using component mode synthesis and large-scale (unreduced) models was used. The bodies featuring moving interfaces were not reduced. Maglie [7] extended the axis construction kit (ACK), presented by Lorenzer et al. [8], for the use with reduced finite element models. For model reduction, the block Arnoldi implementation MOR for ANSYS, presented by Rudnyi and Korvink [9], was used. This tool did not feature any error estimation. Wabner et al. [10] presented a concept for simulation parallel to the design process. A virtual prototype was derived for assistance during design and controller parametrisation. Finite element models were reduced using modal reduction. Uhlmann et al. [11] presented a Kalman-filter, based on a reduced order finite element model used for a high-dynamic controller of a linear axis with flexible support. For model reduction, MOR for ANSYS has been used.

Many applications used to use rigid multi-body simulation. However, in order to account for the elasticity of machine tool structures, it is mandatory to use flexible-multi-body simulation. Even though there are some applications using reduced-order finite element models of machine tool components, none of them use accurate model reduction methods with controlled error bounds. Also, the accuracy of the reduced models is not verified in general.

In order to make the creation of virtual machine tool prototypes applicable, however, a comprehensive framework for efficient and accurate simulation of machine tools is required. The Ph.D. thesis of Spescha [12] is dedicated to this topic and the key methods are outlined in the contribution at hand.

Specific Properties of Machine Tools

General properties of machine tools imply some requirements on software and methods for machine tool simulation. In the following, the key properties of machine tools are listed.

Small strains and deformations—The structure of machine tools is deformable; and it is crucial to model these structural deformations. However, the strains and deformations in structural components are usually small and the structural components can be considered linear-elastic.

Decaying excitation—The excitation of machine tool structures decays with frequency. The structure is excited by, e.g. drives and process forces. The excitation by the drives is limited due to the low-pass characteristics of the controllers and the excitation spectra of process forces decrease with increasing frequency. This allows for the definition of an excitation threshold and thus a relevant frequency range of interest.

Large translations—Moving linear axes undergo large translations. This leads to moving contact points between the individual axes, e.g. at the linear guides, ball screws, or linear direct drives, and therefore, to structural changes of the machine tool. Today, machine tools are usually simulated in one pose only, but the vibration behaviour of a machine tool at changing operation points has also to be taken into account for evaluation of the overall performance. Furthermore, in precision engineering, position dependent static deformation due to static loads, such as gravitation, are also of major interest.

Large rotations—Large rotation of bodies, in contrast to infinitesimal rotations which could be linearised, is inherent in the motion of rotary axes. This leads, similarly to large translations of linear axes, to structural changes of the machine tool.

Distinct interfaces—Interfaces to bodies of machine tools arise at components like bearings, linear guides, motors, gears, measurement systems, or tool-work-piece-interaction. The interfaces to bodies are distinct in terms of location and behaviour. The location of action can change, but is determined by the positions of the axes. The behaviour can be linear or non-linear and both forces and torques have to be applicable just like translations and rotations have to be evaluable.

Mechatronic systems—Due to the integration of structural components, drives, and controllers, machine tools are mechatronic systems. Therefore, a simulation environment has to enable modelling of the properties of those mechanical, electromechanical, and electronic devices by means of a system simulation environment.

Requirements on Simulation Software for Machine Tools

The properties of machine tools impose some requirements on simulation software. Because of the linear-elastic behaviour of structural components of machine tools and thanks to decaying excitation spectra, model order reduction methods for linear, time-invariant systems can be used for the reduction of single components of a multibody system. In order to ensure sufficient accuracy of the reduced models, an accurate estimation of an upper bound for the error of the frequency response functions within the frequency range of interest has to be used. A novel model reduction method featuring error estimation is presented in the discussion on *Model order reduction*.

Considering moving interfaces, a method for the representation of moving interfaces has to be implemented that can be used in combination with model order reduction. Modelling of moving interfaces by interacting with each node individually as described, e.g. by Zaeh and Siedl [6] does lead to a large number of interfaces and is thus not compatible with model order reduction. Therefore, a different approach for modelling moving interfaces has to be used. In the discussion of *Moving interfaces*, a method for modelling of moving interfaces by a minimum number of inputs/outputs is presented.

Due to typically tight time restrictions during development of new machine tools, an effective and straight-forward workflow from the beginning of the modelling procedure until the end of the model analysis is crucial. In the discussion of *Software implementation*, the simulation software MORe is presented. This software implements the presented methods for model order reduction and modelling of moving interfaces within a complete framework for simulation of machine tools.

Model Order Reduction

Some research works in the field of model order reduction featuring error estimation have been published. Fehr [13] presented a fully automated reduction algorithm with iterative selection of expansion points at the frequency with maximum estimated error, automated selection of the number of moments to match, and automated stopping of the Krylov iteration based on tolerances for the change of the estimated error. Grimme [14], Bechtold et al. [15] and Bonin [16] presented heuristic error estimation methods based on complementary systems with different reduction parameters and multiple expansion points for Krylov subspace based reduction. Because machine tool models usually feature a large number of system inputs, Krylov based reduction with multiple expansion points and multiple matching parameters per input usually leads to reduced systems with inadequately high order. Moreover, evaluation of the presented error criteria is a computationally intensive task. Therefore, a model order reduction technique featuring a priori estimation of an error bound has been developed. The method based on projection onto a combination of Krylov and modal subspaces (KMS) has been described in detail by Spescha et al. [17]. The basic idea

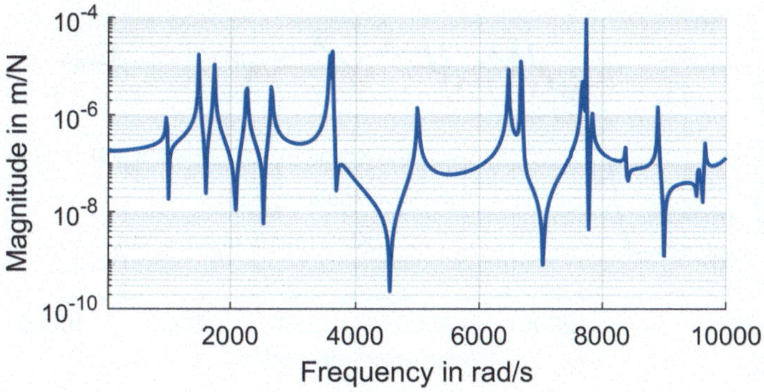

Fig. 1 FRF magnitude of a randomly generated test system

behind the KMS method is to combine the beneficial properties of Krylov subspace based reduction with those of modal reduction.

Krylov subspace based reduction allows matching the transfer functions of reduced and original systems at specific expansion points. In fact, Krylov subspace based reduction leads to Padé approximation of the transfer function. By choosing multiple expansion points, the transfer functions can be matched at multiple Laplace parameter values. However, there is no guarantee that the poles of a transfer function are matched.

On the other hand, modal reduction leads to explicitly matching eigenvalues of a system, i.e. the poles of a transfer function. The overall quality including static compliance, however, is insufficient.

The combination of a Krylov subspace and a modal subspace, i.e. the sum of these subspaces, leads to matching transfer functions at the expansion points for Krylov subspaces as well as matching poles. In order to visualise this, the frequency response functions (FRFs) of a randomly generated system, shown in Fig. 1 and its reduced systems are compared, whereby the resulting order of the reduced system is 10 for each variant. The relative errors are shown in Fig. 2.

The expansion point s_e for the Krylov subspaces is chosen at $s_e = 0$ that leads to numerically exact static matching. For Krylov reduction only, the relative error is very low for a low frequency range but suddenly grows at approximately 700 rad/s.

For modal reduction, 10 modes have been considered, i.e. $n_M = 10$. This leads to exact matching of the first 10 poles. The zeros of the FRFs, however, are not matched exactly, what leads to poles in the relative error FRFs.

For the system reduced by KMS projection with an expansion point at $s_e = 0$ and considering 8 modes in order to get the same total order, the static error is zero. The relative error at low frequencies is higher than for Krylov based reduction. For frequencies above 1450 up to approximately 5000 rad/s, however, the KMS reduced system shows the lowest error. Considering a relative error bound of 5%, the limit frequency for the system reduced by KMS is at 4150 rad/s, that is more than twice as high as the one for the system reduced by Krylov subspace projection only. The

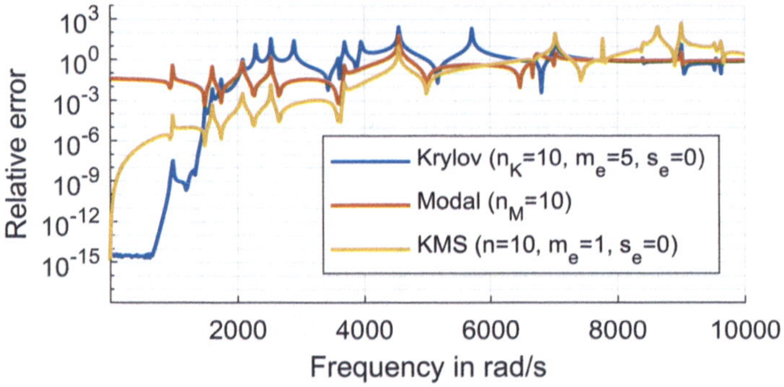

Fig. 2 Relative FRF error magnitudes of the system from Fig. 1, reduced by projection onto a Krylov subspace, a modal subspace, and a KMS

modal only reduction is not feasible, because the requirement on static matching is not fulfilled.

The main achievement with KMS based reduction, however, is not the approximation quality but the good-natured characteristics of the relative error that, as shown by Spescha et al. [17], allows the estimation of a relative error bound, E, according to

$$|E| \leq \prod_{k=1}^{n_e} \left| \frac{\left(\omega^2 - s_{e_k}^2\right)^{2m_{e_k}}}{\left(\omega^2 - \omega_0^2\right)^{2m_{e_k}}} \right|,$$

where n_e is the number of expansion points, s_{e_k} is the kth expansion point, m_{e_k} is the number of Krylov iterations for expansion point s_{e_k}, and ω_0 is the frequency of the first unconsidered mode. An example of the error estimation quality is presented in Fig. 3 for a set of 20 randomly generated systems.

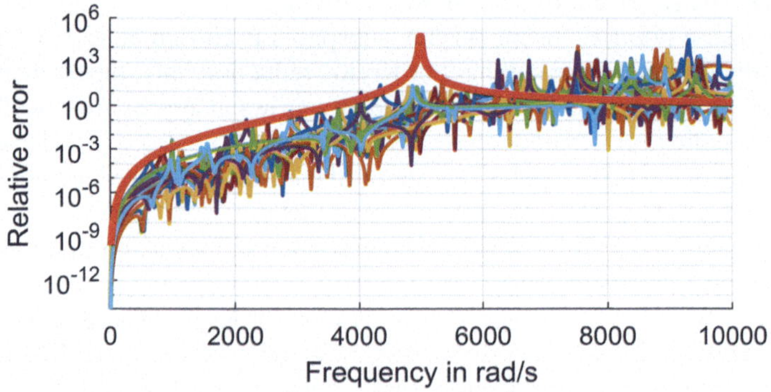

Fig. 3 Estimated error bound (thick line) and relative error FRFs (thin lines) for KMS based reduction of 20 randomly generated systems

Moving Interfaces

In order to model moving interfaces on flexible bodies represented by reduced-order finite element models, a novel approach using trigonometric interpolation of a density function has been presented by Spescha et al. [18].

The key idea is to minimise the number of inputs required for the representation of a density function. Therefore, a moving path is defined, along which the particular terms for a Fourier series are calculated. Figure 4 shows exemplary nodal degree of freedom (DOF) vectors for the constant term and the first three sine and cosine harmonics used for trigonometric interpolation.

These inputs are then considered for model reduction. Afterwards, during simulation, the harmonics can be superposed as a Fourier series in order to obtain the desired distribution along the path, resulting in an approximate distribution as, e.g. shown in Fig. 5.

This leads to a minimum number of individual inputs to the system. However, the resulting number of inputs is still significant if the width of the application is considerably smaller than the length of the path. Therefore, it is crucial to use an efficient model reduction method that is insensitive to the number of inputs.

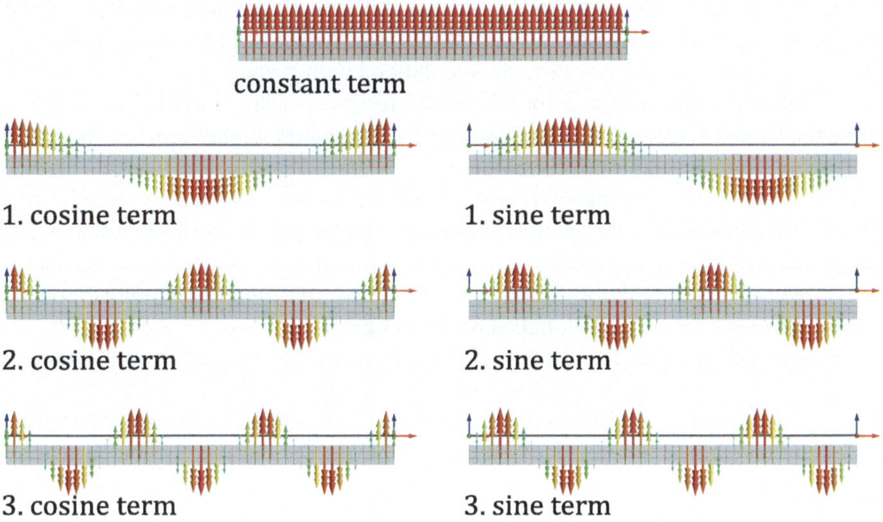

Fig. 4 Nodal DOF vectors for the elementary force distribution in vertical direction. The constant term and the first three harmonics are shown

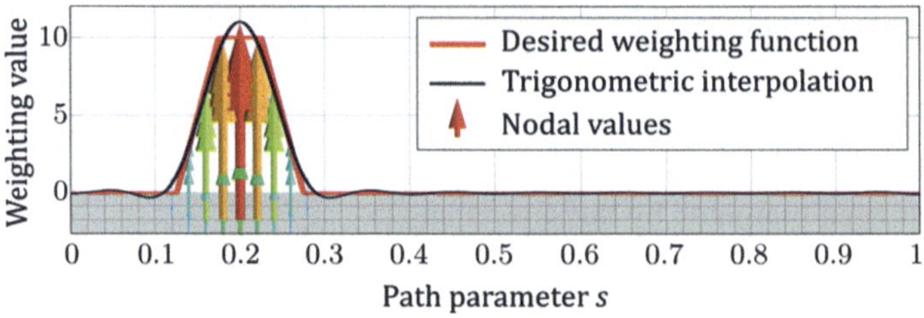

Fig. 5 Nodal DOF vectors and weighting function for trigonometric interpolation of a trapezoidal weighting function

Software Implementation

The simulation software MORe[1] [19], developed at Inspire AG, is designed to comply with the requirements presented in the section *Requirements on simulation software for machine tools*. It implements the KMS based model reduction method presented as well as moving interfaces modelled by trigonometric interpolation.

Special attention has been paid to an effective workflow using the tool chain shown in Fig. 6. Finite element models of single components such as axes, machine bed, or spindles, are created with ANSYS Mechanical without any additional effort, the models are exported by one macro call. After importing a component with MORe, interfaces, either stationary or moving, are defined with graphical assistance. Therefore surfaces are selected and local coordinate systems are specified in order to define the location and orientation of action. Subsequently, model order reduction is performed for each individual component.

The assembly of components to an interconnected composition is realised by creation of links that define the relation between two interfaces or between an interface and the inertial system. By means of specialised link properties, stiffness and damping matrices can efficiently be defined for a multitude of different coupling elements such as linear guides, bearings, ball screws, and gears.

Various analyses can be conducted directly using the graphical user interface of MORe. For the analysis of the mechatronic system, i.e. the controlled structure including drive and controller models, as well as for transient simulation, a Simulink model can be created by means of predefined blocks for the mechanical structure, inputs and outputs to the system, and data management. Finally, for a better insight

Fig. 6 Tool chain of the simulation software MORe

[1] www.more-simulations.ch (accessed 21 May 2019).

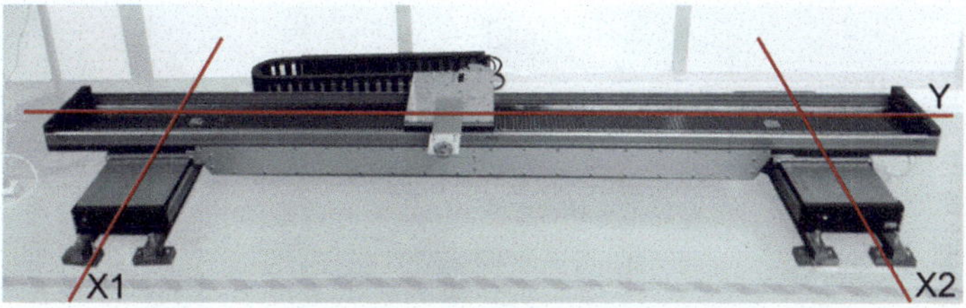

Fig. 7 Gantry stage used for comparison of simulation and measurement results

and interpretation, results from modal analysis, frequency response analysis, and transient simulation can be animated within MORe.

Applications

Methods for creation of comprehensive models of mechatronic systems similar to machine tools have been presented. Such models can be used for versatile analyses for design and diagnostics of new and existing products. In order to illustrate some applications and to show the achievable accuracy, some examples are briefly presented in the following.

Position-Dependent Frequency Response

The dynamic, position-dependent behaviour of a gantry stage (Fig. 7) has been investigated by Lanz et al. [20] by means of position dependent frequency response functions. Due to varying inertia offsets and stiffness while moving the Y-axis, the dynamic properties of the X1- and X2-axis change.

The results for the X1-axis are shown in Fig. 8. Effects, such as resonance and anti-resonance frequency shift or damping variation can be observed in both measurement results and simulation results. Overall, there is good correspondence between measurements and simulation, even though some eigenfrequencies are shifted by up to 9%.

Contour Error

Lanz et al. [20] also present results on transient simulation. The contour error of a rounded cross contour was measured at the encoders as well as at the tool centre point (TCP) for the closed-loop controlled system. The measurement at the TCP was conducted using a Heidenhain KGM grid encoder.

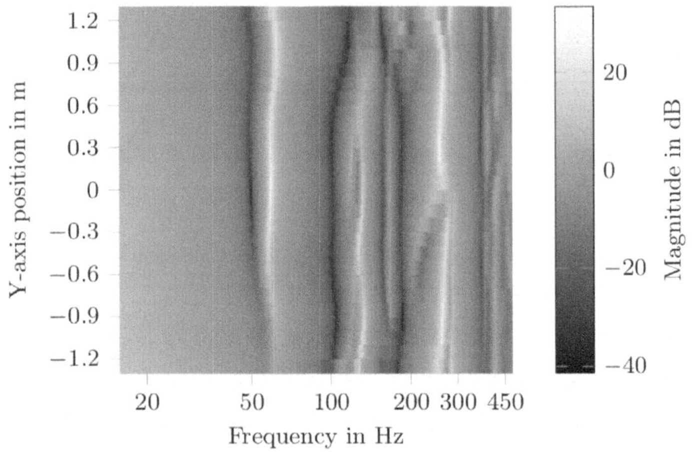

(a) Measurement of the X1-axis

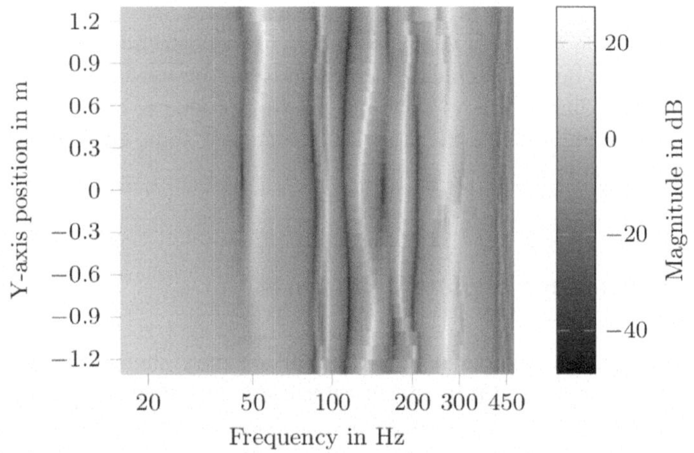

(b) Simulation of the X1-axis

Fig. 8 Comparison of measured and simulated open-loop frequency response functions for varying Y-axis positions of the gantry stage from Fig. 7 (from Lanz et al. [20])

Measurement results are shown in Fig. 9 and simulation results in Fig. 10. There is good correspondence between measurement and simulation results in terms of both quantitative and qualitative comparison. Controller dynamics, structure compliance, as well as backlash due to static friction are well reproduced by the model.

The maximum measured contour error at the TCP is 105 μm and corresponds well with the maximum contour error from simulation, which is 113 μm. The maximum encoder contour errors are 114 μm and 129 μm for measurement and simulation, respectively.

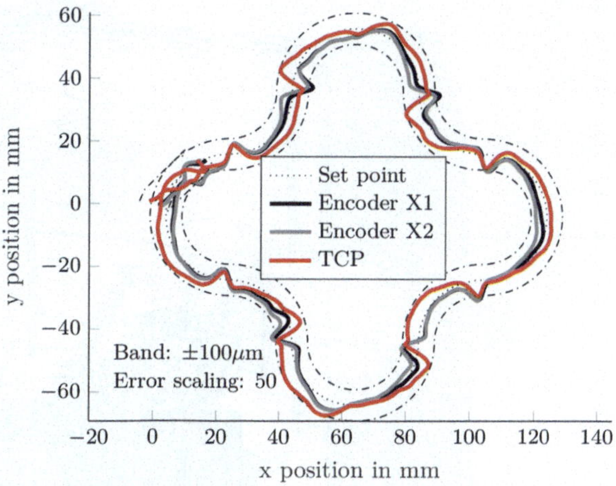

Fig. 9 Measured contour error at the encoders and the TCP

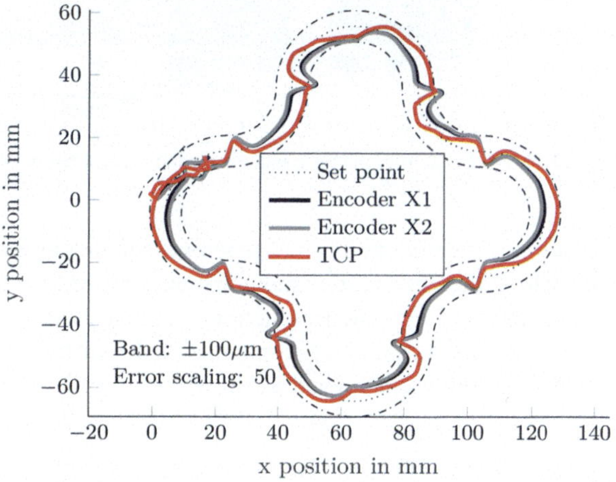

Fig. 10 Simulated contour error at the encoders and the TCP

The simulation time for this contour is approximately 30 s. That is very efficient, although not yet real-time.

Structure and Process Coupling

In manufacturing technology, process stability and productivity tend to be the most vital properties for success. Coupling of a machine tool covering the whole mechatronic system with a production process model represents the top level of analysis. This is only possible with efficient, accurate, and low-order models.

Kuffa [21] showed, by means of a MORe model of a test rig for high performance dry grinding as shown in Fig. 11, that this coupling is feasible. He divided the

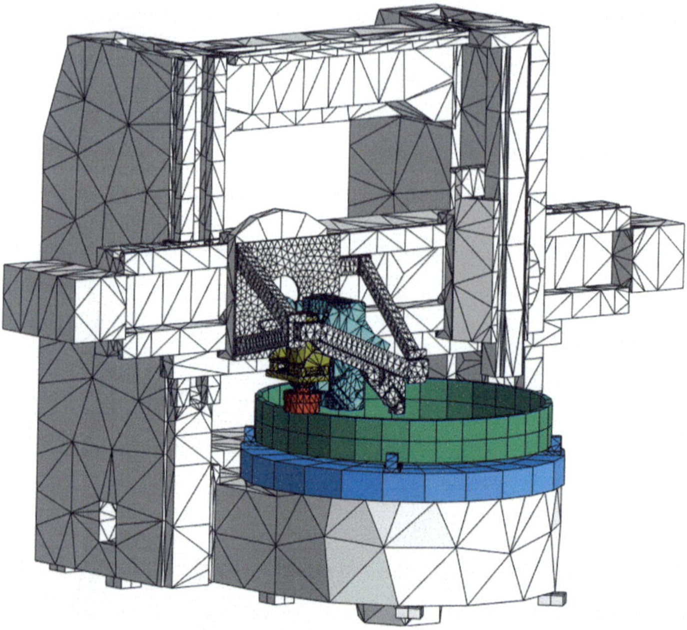

Fig. 11 Model of a high performance dry grinding test rig in MORe (from Kuffa [21])

surface profile calculation in a machine-dependent part and a kinematic roughness part, which depends on the geometry of the grinding grains.

The machine dependent behaviour was analysed by means of a simplified grinding process model. Good correspondence of the surface profile between measurements and simulation was achieved as shown in Fig. 12.

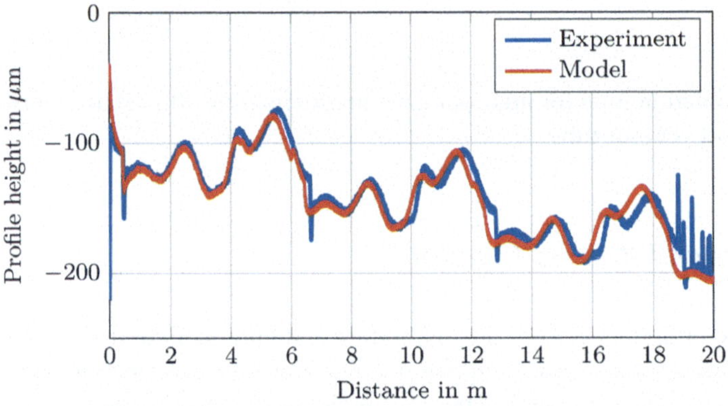

Fig. 12 Simulated and experimental work piece surface profile with descending trend caused by material removal during the grinding process (from Kuffa [21])

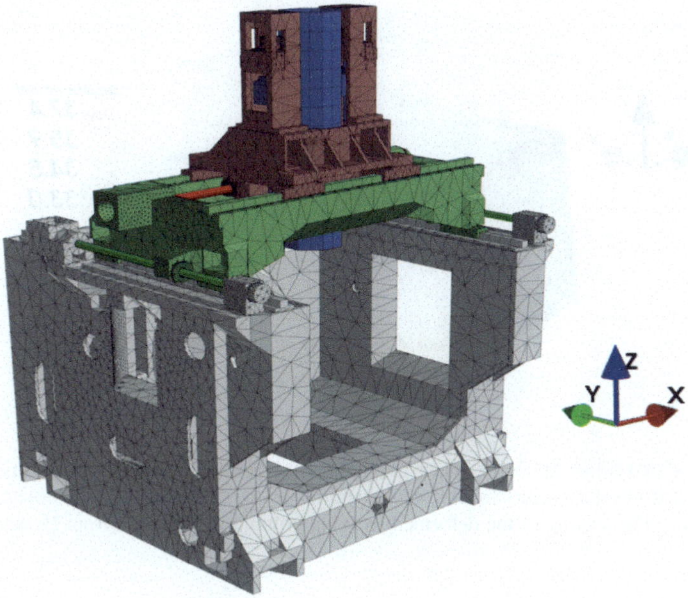

Fig. 13 Model of the Mori Seiki NMV 5000 DCG in MORe (from Hernández et al. [23])

Volumetric Accuracy Evaluation

An important measure in precision machining is the volumetric accuracy of a machine tool. Volumetric accuracy is, according to Ibaraki and Knapp [22], represented by a map of position and orientation error vectors of the tool over the volume of interest.

The calculation of the volumetric accuracy under consideration of static loads, such as gravitation, involves the repetitive calculation of static deformations for an array of axis positions in the workspace.

This is only efficiently possible with a model of low order with matching static stiffness for all possible axis positions, that is achieved with the methods for model order reduction and modelling of moving interfaces presented.

An example of such an evaluation has been presented by Hernández et al. [23] for a real machine, shown in Fig. 13. The total volumetric displacement under gravitational load is shown in Fig. 14.

Conclusions

The simulation software MORe has been presented, which implements methods tailored for the simulation of machine tools with their specific properties.

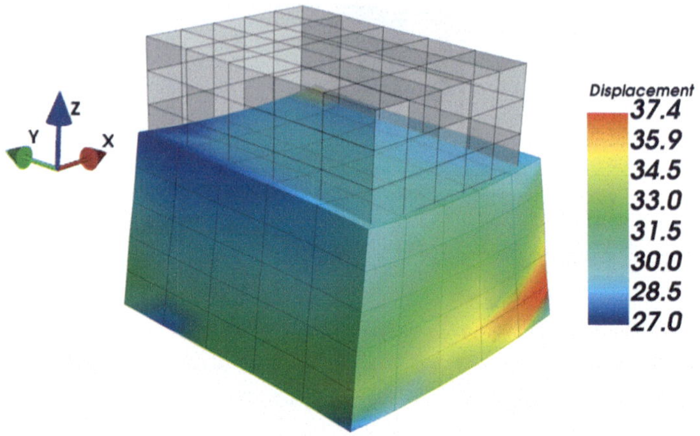

Fig. 14 Simulated volumetric errors due to gravitational forces. Euclidean displacements are shown and color-coded in micrometers for different positions of the X-, Y- and Z- axes over the whole working space. The scaling of the deformed working volume is 10,000:1 (from Hernández et al. [23])

In addition to featuring functionalities for interfacing with finite element software, definition of stationary interfaces, interconnecting and orienting bodies, and interfacing with the system simulation software MATLAB and Simulink, novel methods for model order reduction and representation of moving interfaces are implemented.

The newly developed method for model order reduction based on KMS projection features error estimation a priori to the reduction. That allows a distinct parametrisation without any need for iterations. Thanks to the method for modelling of moving interfaces by means of trigonometric interpolation of density functions, moving interfaces are applicable in combination with an accurate model order reduction method.

Due to its comprehensive functionalities and effective modelling workflow, MORe makes the simulation of the dynamics of machine tools, including drive and controller simulation, applicable in industrial environments. Applications in the field of dynamic performance evaluation and structure-process coupling show promising results.

References

1. Altintas, Y., Brecher, C., Weck, M., & Witt, S. (2005). Virtual machine tool. *CIRP Annals, 54*(2), 115–138.
2. Altintas, Y., Verl, A., Brecher, C., Uriarte, L., & Pritschow, G. (2011). Machine tool feed drives. *CIRP Annals, 60*(2), 779–796.
3. Brecher, C., Esser, M., & Witt, S. (2009). Interaction of manufacturing process and machine tool. *CIRP Annals, 58*(2), 588–607.
4. Weikert, S. (2000). *Beitrag zur Analyse des dynamischen Verhaltens von Werkzeugmaschinen*, ETH Zürich.

5. Berkemer, J. (2003). *Gekoppelte Simulation von Maschinendynamik und Antriebsregelung unter Verwendung linearer Finite Elemente Modelle*. Institut für Statik und Dynamik der Luft- und Raumfahrtkonstruktionen: Universität Stuttgart.
6. Zaeh, M., & Siedl, D. (2007). A new method for simulation of machining performance by integrating finite element and multi-body simulation for machine tools. *CIRP Annals, 56*(1), 383–386.
7. Maglie, P. (2011). *Parallelization of design and simulation: Virtual machine tools in real product development*. Eidgenössische Technische Hochschule ETH Zürich: Institut für Werkzeugmaschinen und Fertigung.
8. Lorenzer, T., Weikert, S., & Wegener, K. (2007). Decision-making aid for the design of reconfigurable machine tools, In *Proceedings of 2nd International Conference Changeable, Agile, Reconfigurable& Virtual Production*, pp. 720–729.
9. Rudnyi, E. B., & Korvink, J. G. (2006). Model order reduction for large scale engineering models developed in ANSYS. In J. Dongarra, K. Madsen, & J. Waśniewski, (Eds.), *Proceedings 7th International WorkshopApplied Parallel Computing. State of the Art in Scientific Computing, (PARA 2004), Revised Selected Papers*, (pp. 349–356). Springer.
10. Wabner, M., Frieß, U., Hofmann, S., Hellmich, A., & Quellmalz, J. (2014). Optimierte Inbetriebnahme durch Simulation, *wt Werkstattstech. online, 1/2*, 97–99.
11. Uhlmann, E., Essmann, J., & Wintering, J.-H. (2012). Design- and control-concept for compliant machine tools based on controller integrated models. *CIRP Annals, 61*(1), 347–350.
12. Spescha, D. (2018). *Framework for Efficient and Accurate Simulation of the Dynamics of Machine Tools*. Faculty of Mathematics/Computer Science and Mechanical Engineering, Clausthal University of Technology.
13. Fehr, J. (2011). *Automated and error controlled model reduction in elastic multibody systems*. Institut für Technische und Numerische Mechanik: Universität Stuttgart.
14. Grimme, E. J. (1997). Krylov projection methods for model reduction. University of Illinois at Urbana-Champaign.
15. Bechtold, T., Rudnyi, E. B., & Korvink, J. G. (2004). Error estimation for Arnoldi-based model order reduction of MEMS. *System, 10,* 15.
16. Bonin, T., Faßbender, H., Soppa, A., & Zaeh, M. (2016). A fully adaptive rational global Arnoldi method for the model-order reduction of second-order MIMO systems with proportional damping. *Mathematics & Computing in Simulation, 122,* 1–19.
17. Spescha, D., Weikert, S., Retka, S., & Wegener, K. (2018). Krylov and modal subspace based model order reduction with A-priori error estimation. *ETH Zurich Research Collection* @ www.research-collection.ethz.ch/bitstream/handle/20.500.11850/284435/20180824_modelreduction.pdf?sequence=1. Accessed May 21, 2019.
18. Spescha, D., Weikert. S., & Wegener, K. (2018). Modelling of moving interfaces for reduced-order finite element models using trigonometric interpolation, *ETH Zurich Research Collection* @ www.research-collection.ethz.ch/bitstream/handle/20.500.11850/284509/1/20180824_moving_interfaces.pdf. Accessed May 21, 2019.
19. inspire AG. (2019). MORe @ www.more-simulations.ch/index.php/contact/. Accessed May 21, 2019.
20. Lanz, N., Spescha, D., Weikert, S., & Wegener, K. (2018). Efficient static and dynamic modelling of machine structures with large linear motions. *International Journal of Automation Technology, 12*(5), 622–630.
21. Kuffa, M. (2017). High performance dry grinding. *HPDG*, ETH Zurich.
22. Ibaraki, S., & Knapp, W. (2012). Indirect measurement of volumetric accuracy for three-axis and five-axis machine tools. *A Review, International Journal of Automation Technology*, *6*(2), 110–124.
23. Hernández, P., Zimmermann, N., Pavlicek, F., Blaser, P., Knapp, W., Mayr, J., & Wegener, K. (2018). Learning efficient modelling and compensation for thermal behaviour of machine tools. *MTTRF 2018 Annual Meeting*.

Reinventing Mechatronics—Final Thoughts

David Russell

Abstract The chapters in this book were selected from papers presented at the 16th Biannual Mechatronics Forum Conference held at the University of Strathclyde in Glasgow in September 2018 on the basis of current interests and are by no means to be construed as an exhaustive work. However, the underlying principles in each chapter are very similar and important. Each points to the future and how the reinvention of many aspects of mechatronic systems is critical if only to avoid repeating the same mistakes as previously made by others. This concluding chapter is therefore deliberately not overly technical, with just a few references to encourage interested readers to delve deeper into specific areas of reinvention.

Introduction

The designers of mechatronic systems of all varieties and types such as those described in previous chapters of this book must be aware of a group of fundamental principles if success is to be ensured. Each principle points to the future in support of how the reinvention of mechatronic systems is critical if only to avoid repeating the same mistakes as previously made by others. This concluding chapter presents a non-exhaustive list of those features including: technology; complexity; operating software; hardware; reliable and rechargeable power supplies; innocent human error; connectivity; privacy, dependency, ubiquity and the hybrid society. Each section is deliberately not overly technical with just a few references to encourage interested readers to delve deeper into specific areas of reinvention.

D. Russell (✉)
Penn State Great Valley, Malvern, USA
e-mail: drussell@psu.edu

X.-T. Yan et al. (eds.), *Reinventing Mechatronics*,
https://doi.org/10.1007/978-3-030-29131-0_12

Technology

Technology companies continue to be focussed on attracting a larger market share of an eager, *probab*ly younger, clientele who are not really cognisant of how the technology that they are using works, but just the utility of it. Getting a new technology accepted is now really a matter of marketing and media exposure. Going back twenty-five years it would have been inconceivable that one could not only monitor visually one's domicile from another country but also switch lights on and off, reset alarms and water the lawn! Those same systems today can order groceries when refrigerator supplies are low, alert emergency vehicles if the homeowner is not moving for some pre-set time period, unlock the front door to admit the medical personnel after supplying vital signs to the ambulance. People check their bank balances, pay bills, deposit cheques and make investment transactions while sitting on a train or having lunch. It would seem that ubiquitous reliable access to technology-enhanced life services is to be expected and always available.

Mechatronic systems may be large, such as a commercial jetliner, or small like an artificial heart. There is certainly no lack of new inventions that could be classified as being of a mechatronic nature, many of which are designed using past methodologies and have the same inherent flaws and real risks of failure. Sometimes systems fail for reasons of complexity, software, hardware, power supply, human error, or connectivity. Incidental societal effects, such as privacy or dependency, continue to happen usually without warning. All these factors must be considered in every design process. The following sections are an attempt to address these aspects and offer commentary as to where reinvention is an urgent necessity.

Complexity

Designers of systems are systems engineers, computer scientists are seldom experienced in the application domain in which they are employed. How a mechatronic system is designed and how it finally interfaces with its environment is a very complex matter, especially in large systems. The crash of two Boeing *737 MAX 8* commercial aircraft in 2019 is a sad reminder of this and will be used in illustration. Without much warning both of these aircraft began pitching erratically soon after take-off and nose-dived to earth killing all aboard. Both aircraft had installed a new safety system that was designed to alter the pitch angle of the plane to avoid stalling based on sensors in the nose of the aircraft. Apparently the pilots could have disabled it but were confused and possibly not completely trained in the new system. In both instances, the sensors reported erroneous readings causing the anti-stall system to kick in and drive the aircraft into the ground.

In March 2019, the *Seattle Times* [1] reported that Boeing was finalizing its proposed software fix for the *737 MAX 8*, with flight tests likely to begin shortly afterwards and that "*... the software patch will be provided to airlines* 'free of charge' *once* *it has been certified by the FAA and released*." The *Seattle Times* further explains that the software patch will revamp how the *Manoeuvring Characteristics Augmentation System* works, including using input from both of the *737 MAX 8*'s angle of attack sensors and ensuring that the anti-stall system is not triggered multiple times, as it was in the *Lion Air* crash. In a subsequent article [2] the *Seattle Times* reported that the company said it will only take about "*an hour or so*" to upgrade an airplane with the new software.

Future systems must include the ultimate users in all stages of specification, design and proving. The end-user must be able to accommodate the mechatronic system easily and know how to escape from it talking back a safe manual control. Along the same lines, any documentation or on-line help materials should be available in other languages as a configuration option. Systems must be able to evaluate the validity or quality of signal from a sensor and be able to distinguish erroneous or missing data, and react accordingly.

Operating Software

In addition to complexity and unforeseen system situations as outlined above in which the control software was flawed, there are other software areas where problems occur. These issues have plagued mechatronic and indeed all commercially available systems. When a system is revised or repaired incompatibility with the operating system or system software can inadvertently become present. As better and faster chip sets with their improved assembler instructions are introduced into a product, the software designer in previous revisions may depend on access to prior instructions causing sections of the code to fail when in situ. This is particularly important in real-time software where machine level interrupts drive a system between modules and sub-systems. Was the repeated entry and exit into the *737 MAX 8* stall prevention system caused by such a problem?

Operating system and software upgrades are a necessary part of technological advances but future systems must accommodate change and only be allowed after rigorous dynamic testing. In real-time systems, each action's software is usually memory resident and quite small in size. These agents not only perform their own set tasks but cooperate in overall system flows. This includes the execution of what are known as "*finite state machine*" entry and exit semaphores, interrupt processing which may require hardware flag access, the avoidance and detection of any unplanned logic loops and the safe handling of all errors. The software must be reconfigurable on-the-fly and be easily restarted on recovery. Of course, once the system is an integral part of a distributed network with anonymously written code provided by from software as a service mechanism, these limitations can only be investigated by almost infinite testing.

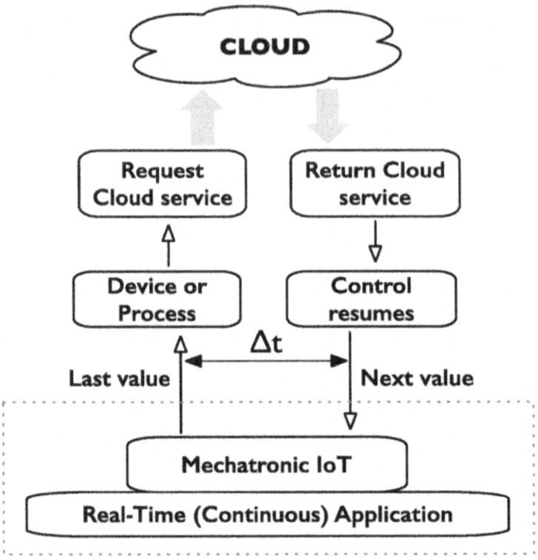

Fig. 1 IoT devices and the cloud

In any real-time mechatronic system, the prevailing issue is that of latency. Figure 1 illustrates how devices that are classifiable as being Internet of Things (IoT) functional access cloud services be they data, software-as-a-service, platform-as-a- service, or system-as-a-service.

These same devices inherit the same uncertainty and risk factors present in any internet-based system. Figure 2 illustrates this. The software designer must accommodate the possibility of fairly long outages and the possibility of varying sample rates and irregular interrupts. For example, when a hard deadline is reached, the device software, which may be embedded deep in the system, must always return a safe and valid response, which while preferably obtained from the cloud may be a pre-set default value.

Hardware

Any physical component that is released to the general public will have reliability and eventual failure issues and will need replacement. These emergencies may be due to loose wiring and joints, component wear and tear, and accidental damage by the user. This is critical especially in the medical arena where devices are comparatively small. For example, in March 2015 BiVACOR® [3] announced what they called the world's first truly bionic heart. The team had successfully implanted it in a sheep, and human trials are expected to begin in 2018. It is anticipated that the BiVACOR heart will last ten years and be smaller and more reliable than any artificial heart ever built. A ten or even twenty year replacement cycle is not really long. The term *device fatigue* is a really a euphemism for possible and perhaps even expected failure.

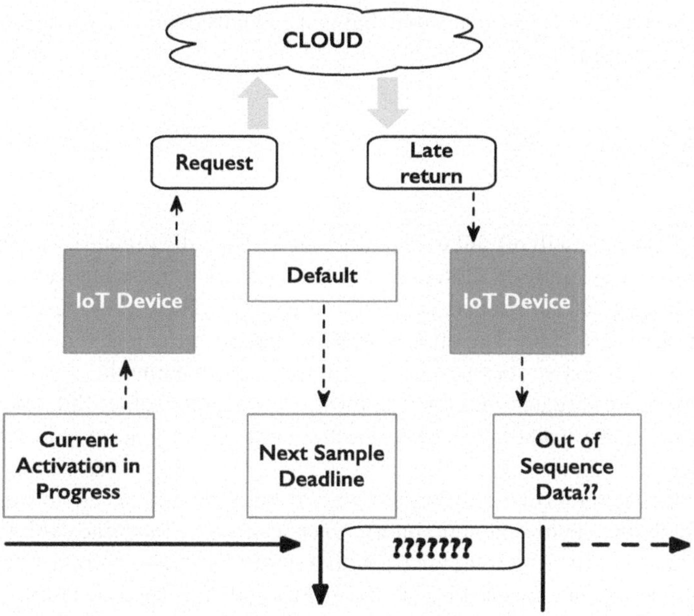

Fig. 2 Cloud latency and timing uncertainty for system as depicted in Fig. 1

Reinvention is needed in systematic reviews of emerging technologies especially if the new device or processor, while appearing to be more powerful or reliable, retains all of the features at the chip level of its predecessor. In a biomedical application, tissue rejection or device deformation may cause hitherto unforeseen problems.

Reliable and Rechargeable Power Supplies

All mechatronic systems and components generally require an energy source be it chemical, solar or electrical. In addition to an actual charging mechanism, which may be non-wired contact, the user or system must recognise the need for recharging and indicate forcefully of that immediate need. An eventual swap out of the power unit will be necessary at some unknown future time. For example, the BiVACOR artificial heart mentioned above requires an external unit with batteries that provide a variable current draw based on particular smart and adaptive functions that the mechatronic heart supports in response to the recipients activity. This implies that the mechatronic systems of the future must be smart and active rather than the more customary passive assistive devices. The power supply and management system must now be an active component in any design and include active operations should a fault or power starvation occur. This may have to go beyond warnings and include redundant system backup.

In large systems this is quite normal, but in a paediatric implant on the other hand space is at a premium.

Innocent Human Error

Devices and systems will all fail eventually, but perhaps the greatest source of malfunction is the innocent user. Casual or random operation, for whatever reason and under no particular circumstance, may drive the system into a previously uncharted unstable or erroneous state. The effects of malicious misuse can never be completely eradicated from a system, but loopholes in the design that enable casual entry into what may be a life threatening situation must never be allowable. Software designers should not leave backdoor access into the code. Clearly some precautions are common sense.

As an example of this in the author's experience was the easy access into a machine room in which the electromagnetic interference, caused by machine cycling, was so powerful that it was deadly to anybody with a pacemaker. The entrance was posted with a sign but was not secured by an access key coded to exclude employees with such devices. In an example from a more sophisticated environment, an aircraft pilot must be able to disable the auto-pilot and fly manually under any circumstances and at any speed. The *737 MAX 8* disaster report indicated that the manual operation of the plane was not humanly possible due to the high speed of the aircraft causing so much force on the tail unit that it could not be controlled. A fly-by-wire setup may have enabled manual tail control, but only if designed to handle high speed forces.

It is clear that proper documentation, instruction, and care of any mechatronic device is of paramount importance and that instruction manuals be written clearly in simple language to enable novice users to operate the system even under system or component failures.

Connectivity

Perhaps the most trusted and yet least understood factor of technology is the provision of reliable, secure, and robust interconnection between users, systems, and systems of systems which include The Cloud. End users must depend heavily on internet service providers (ISP). Should a system fail, rapid restoration of service must be available, and yet the user has no control over the process. Commercial IT companies are factoring digital disruption into their strategic operation plans and purchasing temporary backup facilities. These backup sites cover failures in power and network, physical disasters, and the ever present looming spectre of terrorism. In biomedical or other dynamic systems, this luxury cannot exist so in situ remedies such as default value processing, external batteries or even front-door keys must still work.

Privacy, Dependency, Ubiquity and the Hybrid Society

There is a real acceptance of automated systems in the domestic, commercial, and engineering world. Passengers on a flight between two airfields expect the auto-pilot and landing systems to essentially fly the aircraft safely based on the coordinates being provided dynamically as the flight progresses. On all but the smallest of aircraft, this is a de facto dependency. Using a smart phone to meter out exercise regimens or to order food delivery is almost passé. Implanted medical devices are increasingly able to restore the quality of life to stroke sufferers etc. Portals allow the general public to view their medical history and test results and to make future appointments. Yet there is a hue and cry when such data is accessed or corrupted by nefarious hackers. Even mega companies such as British Airways and Facebook are susceptible to password and personal data breaches [4] and in which millions of records were captured for resale to telemarketing companies.

Looking at current mechatronic devices and the promise of massive increases in the integration of smart technology into society, it is apparent that not all actors will be as agreeable as others. For example, in a smart city all emergency vehicles must be equipped with traffic light control systems and all traffic and railway signals must be equally compliant. Driverless cars will need to be programmed to accommodate reckless driving practices by possibly impaired human drivers. The future society will indeed always be hybrid as long as automation is an optional and initially expensive luxury addition. Persons will always demand the right to opt out or opt into almost all automated systems, to require service availability on all devices, and to be assured that their use is private [5] and their data secure. Much of this must be imbedded in the fabric of the design of mechatronic, and in fact all, systems.

Final Thoughts

As this volume concludes the reader should reflect on how technological systems have positively advanced society over a relatively short time period and yet do have an unfortunately chequered past of failures and disasters. As mechatronic systems become more ubiquitous, so worldwide cultures have expectations far beyond technical specifications and novelty. It is time to reinvent the design, implementation and availability of mechatronic systems to meet those expectations.

References

1. www.seattletimes.com/business/boeing-aerospace/boeing-has-737-max-software-fix-ready-for-airlines-as-dot-launches-new-scrutiny-of-entire-faa-certification-process/. Accessed 25 Mar 2019.

2. www.seattletimes.com/business/boeing-aerospace/boeing-details-its-fix-for-the-737-max-but-defends-the-original-design/. Accessed 27 Mar 2019.
3. Russell, D. W. (2018). Mechatronics and the Internet of Things: A Solution or a Problem? In: *Proceedings 16th Biannual Mechatronics Forum Conf*erence, University of Strathclyde, (pp. 10–16).
4. Data Breaches at www.google.com/search?client=safari&hl=en-us&q=data+breaches+list&sa=X&ved=2ahUKEwirh7LH5c3hAhXLtlkKHfjyAvMQ1QIoA3oECA4QBA&biw=768&bih=985. Accessed 12 Apr 2019.
5. Watt, S., Milne, C., Bradley, D. A., Russell, D. W., Hehenberger, P., & Azorin-Lopez, J. (2016). Privacy matters—issues within mechatronics. *IFAC-PapersOnLine, 49*(21), 423–430.

The manufacturer's authorised representative in the EU is Springer
Nature Customer Service Centre GmbH, Europaplatz 3, 69115 Heidelberg,
Germany. If you have any concerns regarding our products, please
contact ProductSafety@springernature.com

Printed and bound by CPI Group (UK) Ltd, Croydon, CR0 4YY
30/09/2026
02225458-0004